Homogenization and Structural Topology Optimization

Springer
London
Berlin
Heidelberg
New York
Barcelona
Hong Kong
Milan
Paris
Santa Clara
Singapore
Tokyo

Behrooz Hassani and Ernest Hinton

Homogenization and Structural Topology Optimization

Theory, Practice and Software

With 144 Figures

Springer

Behrooz Hassani, MSc, PhD
Shahroud University, Shahroud 36155-316, Iran

Ernest Hinton, BSc, MSc, PhD, DSc, CEng, MIStructE, MBCS
Department of Civil Engineering, University of Wales, Singleton Park, Swansea, SA2 8PP

Additional material to this book can be downloaded from http://extras.springer.com.

ISBN 978-1-4471-1229-7 ISBN 978-1-4471-0891-7 (eBook)
DOI 10.1007/978-1-4471-0891-7
British Library Cataloguing in Publication Data
Hassani, Behrooz
Homogenization and structural topology optimization : theory, practice and software
1. Structural optimization 2. Topology 3. Structural engineering 4. Homogenization (Differential equations)
I. Title II. Hinton, E. (Ernest)
624.1'7713
ISBN 978-1-4471-1229-7

Library of Congress Cataloging-in-Publication Data
Hassani, Behrooz, 1960-
Homogenization and structural topology optimization : theory, practice and software / Behrooz Hassani and Ernest Hinton.
p. cm.
Includes bibliographical references and index..
ISBN 978-1-4471-1229-7 (Berlin : acid-free paper)
1. Structural optimization. 2. Topology. 4. Homogenization (Differential equations) I. Hinton, E. (Ernest) II. Title.
TA658.8.H37 1998 97-49066
624.1'7713'0151 — dc21 CIP

Softcover reprint of the hardcover 1st edition 1999

Typesetting: Camera ready by authors

69/3830-543210 Printed on acid-free paper

In memory of our fathers

Ghorban Ali Hassani and Stanley G. Hinton

Preface

Objectives — In recent years structural topology optimization has found its way into industry and is now used in a variety of engineering fields. Despite considerable research in this subject relatively few books have been published to date. The main aim of this introductory book is to assist the interested reader with an engineering background to become familiar with the concepts, theories and applications of the subject.

This book is intended for advanced undergraduate and graduate students of civil, mechanical and aeronautic engineering and practicing engineers and designers in the aerospace, marine and automotive industries. It may also be of interest to applied mathematicians and architects. The software is provided as source code on the CD-ROM and can be used as a starting point by new researchers in the fields of structural topology optimization and homogenization. The software included can also be used by teachers of basic structural engineering courses to strengthen the understanding of students of load carrying systems.

There are different approaches to structural topology optimization which vary from intuitive methods such as hard kill/soft kill to more mathematically rigorous ones such as the so called homogenization method which is the emphasis of this book. Homogenization is a mathematical theory with several applications in engineering problems which are defined on domains with regular microscopic heterogenities. Artificial material models can be used in place of homogenization theory in structural topology optimization. These material models are simple and usually the resulting layouts are of a more practical nature. However, when the artificial material models are used, the values of the objective functions become distorted. Homogenization can be employed for the evaluation of results from such models and to compare different solutions.

Background — Having described the objectives, EH now provides some background history to the book's development: 'About 10 years ago, after working on various aspects of finite element simulation for 20 years, I decided to re-focus my research, concentrate on structural and multidisciplinary optimization and set up a research group on adaptive design optimization called the ADOPT Research Group. Optimization was a topic I had been interested in since my student days and the first course I taught at Swansea was one called 'Civil Engineering Systems' and contained much on optimization techniques. Indeed, as a result of teaching this course, I subsequently co-authored a textbook entitled '*Civil Engineering Systems Analysis and Design*' with Alan Smith and Roland Lewis which was published by John Wiley in 1984 and which is still in print.

'By this time, two highly successful international conferences had been held at Swansea featuring structural optimization as the main theme. These had been organized by Professor Olek Zienkiewicz who was the Head of the Civil Engineering Department. These meetings had further stimulated my interest in this topic. Unfortunately, at that time computers were not so powerful and accessible, finite element simulation tools were not as well developed and optimization algorithms were still being refined and improved.

'Several years later, Olek drew my attention to the pioneering work of Martin Bendsøe and Noburu Kikuchi on topology optimization using homogenization concepts. I found their early work fascinating and vowed to myself that I would investigate it further at the earliest opportunity. So began an interesting voyage of discovery and when Behrooz Hassani knocked on my office door at Swansea to discuss potential research topics for his doctoral studies, I tried to encourage him to work with me on this exciting new research area. This book is the direct result of our collaboration over the years in trying to come to grips with various aspects of topology optimization.

'Having successfully completed his PhD studies and one further period of postdoctoral research at Swansea, Behrooz went to work for Altair Inc., an American software company with many international offices, who market a commercial code for topology optimization called OPTISTRUCT founded on concepts similar to those discussed in this book. This book was almost completed and existed in draft form by the time that Behrooz left Swansea to join Altair at their Newport Beach offices in California. The book has now been completed after many sessions with TeX and with the help of ADOPT Research Group member Simon Bulman who checked and updated the codes and test data included on the CD-ROM with the book'.

Layout and use of the book —The book has been divided into three parts. Part I is devoted to the homogenization theory where the derivation and solution of the equations for the material models used in topology optimization are presented. The reader who is not interested in homogenization may study Section 3.2 only and skip the rest of Part I. In Part II an algorithm for structural topology optimization based on optimality criteria methods is developed and its practical applications are discussed. Part III is devoted to the introduction of a few alternative methods to structural topology optimization. The concepts of integrated structural optimization are also presented to provide the reader with an overview of the latest developments in the field.

We hope that readers will find the book stimulating and helpful in trying understand topology optimization and we see our text as complementary to the excellent ground-breaking textbook, '*Optimization of Structural Topology, Shape and Material*', written by Martin Bendsøe and also published by Springer Verlag which we strongly recommend you to read. For another viewpoint on topology optimization we also recommend the most stimulating book, '*Evolutionary Structural Optimization*', by Mike Xie and Grant Steven again published by Springer.

CD-ROM — Included with the book is a CD-ROM with two programs: HOMOG and PLATO. HOMOG is a program that may be used to evaluate the homogenized constitutive coefficients for material idealisations to be used under plane stress assumptions. The material to be homogenized consists of a set of repeated identical microcells. An important constraint of HOMOG is that each microcell must have bilateral symmetry. User instructions may be found in Appendix D together with some examples in the form of benchmarks. PLATO is a program that may be used to find optimal structural topologies using homogenization-type solutions based on bi-cubic and bi-quartic models as well as artificial material models described in this text. User instructions for PLATO may be found in Appendix E together with some benchmark examples. These codes are restricted to idealisations based on 2D plane stress assumptions.

Both programs are written in Fortran 77 and use the finite element method as the stress analysis simulation tool. Although every attempt has been made to verify the programs, no responsibility can be accepted for their performance in practice.

Finally, we invite constructive comments and suggestions from readers on the text as well as the codes and examples included in this book.

Behrooz Hassani
Ernest Hinton

Swansea, September 1998

Acknowledgements

It is our pleasure to thank friends and colleagues at the University of Wales Swansea including Dr. J. Sienz, Dr. J. Bonet, Dr. D. Peric, Dr. B. Boroumand, Dr. B. Koosha, Dr. M.T. Manzari, Dr. S.J. Lee and the many members of the ADOPT Research Group past and present for their friendship, support and useful technical discussions.

We also thank Prof. O.C. Zienkiewicz, Prof. P. Bettess, Prof. G. Steven, Prof. A.R. Díaz, Prof. E. Ramm, Dr. O. Sigmund, Dr K. Maute, Prof. K.U. Bletzinger, Dr. J. Bull, Prof. M. Papadrakakis, Dr. M. Xie, Dr. V.V. Toropov, Prof. G. Thierauf and Dr. F. van Keulen for useful communications during our research work.

Thanks are also due to BH's colleagues at Altair Computing Inc, Dr. H. Thomas, Dr. N. Pagaldipti, Mr. B. Vote and Dr. Y.K. Shyy for their assistance.

Special thanks are due to Karen E. Wicks for her editorial assistance, to Dr. A. Jalali-Naini for his advice related to some mathematical aspects of this work, to Simon Bulman for his considerable assistance in updating the programs included in this book and to Adrian Hooper, Andrew Lennon and Euan Wood for help in running the various benchmarks.

This book has been written in TEXand would not have been completed this millennium without the TEX wizardry and typographical cunning of Hans Sienz.

The support of EPSRC U.K. funding agency, Shahroud University and Behim Dezh Company (Iran) as well as the stimulating forums provided by ISSMO, NAFEMS and IACM are also gratefully acknowledged.

We are happy to acknowledge the pioneering work of Martin Bendsøe and Noboru Kikuchi and their collaborators which greatly stimulated and influenced our efforts.

EH would also like to thank Dr. P.P. Strona and staff at the Fiat Research Centre (CRF), Turin, Italy for their helpful collaboration.

Finally, our special thanks to our families for their encouragement and moral support.

Behrooz Hassani
Ernest Hinton

Swansea, September 1998

Contents

Part II: Topology Optimization

4. Structural Topology Optimization using Optimality Critieria Methods

5. Experiences in Topology Optimization of Plane Stress Problems

8. Integrated Structural Optimization

Notation

Abbreviations:

BCs	Boundary Conditions
CAGD	Computer Aided Geomertic Design
CAO	Computer Aided Optimization
DCOC	Discretized Continuum-based Optimality Criteria
DOT	Design Optimization Tools: commercial code
FDM	Finite Difference Method
FE	Finite Elements
HOMOG	Program for evaluating **HOMOG**enized elasticity coefficients
MMA	Method of Moving Asymptotes
MR	Mindlin Reissner
NAFEMS	National Agency for Finite Element Method and Standards
MBB	Messerchmitt-Bölkow-Blohm
ndof	number of degrees of freedom
OC	Optimality Criteria
PLATO	**PLA**ne Stress Topology Optimization Program
R1	Rank-1 material model
R2	Rank-2 material model
SAM	Semi Analytical Method
SCB	Short Cantilever Beam
SE	Strain Energy
SIMP	Solid Isotropic Microstructure with Penalty
SKO	Soft Kill Optimization method
SQP	Sequential Quadratic Programming
SO	Shape Optimization
SSO	Structural Shape Optimization
2D	Two Dimensional
3D	Three Dimensional

Scalars, Functions and Tensors

a	hole side length dimension of microcell in homogenization model
	principal radius of generalized ellipsoid
	constant of integration
a_i	share of Gauss point i in the volume of element
$a(\mathbf{u}, v)$	energy bilinear form for the internal work

a_i	cross-sectional area of truss member i
a_i^o	lower bound of the cross-sectional area of truss member i
a_{ij}	coefficient of matrix $\mathbf{A}$
	functions which are periodic in the spatial variables $\mathbf{y} = (y_1, y_2, y_3)$
A	area
A_{all}	allowable area
b	hole side length dimension of microcell in homogenization model
	constant of integration
	principal radius of generalized ellipsoid
	width of cross section of beam
	filter factor in convolution method
c	principal radius of generalized ellipsoid
c_k	coefficients of approximating polynomial
c_{ijkl}	component of fourth order constitutive tensor
$C(0), C(2)$	order of continuity
C_1, C_2	constant
$C_{ij}, (i,j = 1,\ldots,6)$	elements of matrix of rigidities
d	displacement
d_{all}	allowable displacement
dA	infinitesimal area
$d\ell$	differential arc length
$D_{ij}, (i,j = 1,2,6)$	elements of matrix of elastic rigidities in compact matrix form
D_{ij}^{I}	elements of elasticity matrix for strong layer of R1 material
D_{ij}^{II}	elements of elasticity matrix for weak layer of R1 material
D_{ij}^{R1}	elements of elasticity matrix for rank-1 material
$D_{ij\,m}^{H}$	elements of homogenized membrane rigidity matrix
e_{kl}	component of second order strain tensor
$D_p^n(a,b)$	complete polynomial of degree n
$\hat{D}(a_i, b_j)$	elements of D matrix at sampling points
E	elastic modulus
$E_1,\ E_2$	elastic moduli in orthotropic material
E^H	homogenized modulus of elasticity
E_i	elastic modulus of truss member i
E^ϵ	Young's modulus in homogenization
$E_{ijkl}^H(a,b)$	homogenized fourth order constitutive tensor
$E_{ijkl}^G(a,b,\theta)$	general elasticity tensor
E_{min}	very small elastic modulus for the removed

	elements
EI	flexural or bending rigidity
f	internal heat generation
f_i	internal force in truss member i
f_i^ϵ	body force in homogenization of elasticity problem
$f(\mathbf{x})$, $F(\mathbf{x})$, $F(\mathbf{s})$	objection function
$\mathcal{F}$	general function
g	acceleration due to gravity
	general function
g^σ	stress constraint
g^d	displacement constraint
g^v	volume constraint
$g_k(\mathbf{x})$, $g_j(\mathbf{s})$	inequality constraint function
h_i	local coordinate of point i
$h_k(\mathbf{s})$, $h_j(\mathbf{x})$	equality constraint function
I^e	total potential energy from element e
J_1, J_2	terms in homogenized constitutive matrix for rank-1 material
$\lvert\mathbf{J}\rvert$	determinant of the Jacobian matrix
k	iteration number
	a weighting factor
$K^\epsilon(x)$	conductivity coefficient in homogenization
K^H	homogenized conductivity coefficient
L	length of bar
	span of beam
ℓ_i	length of truss member i
$\ell(\mathbf{u})$	mean compliance
$\mathcal{L}(\mathbf{x}, \mathbf{s}, \boldsymbol{\lambda}, \boldsymbol{\nu})$	Lagrangian function
m	number of elements
M_x, M_y, M_{xy}	bending moments
n	number of quantities
	number of nodes
	number of load cases
n_1, n_2, n_3	arbitrary integer numbers
n_{dv}	number of design variables
n_e	number of elements
n_H^e	number of neighbouring elements
n_g	number of inequality constraints
n_{ga}	number of active inequality constraints
n_h	number of equality constraints
n_n	number of nodes
np	number of points (nodes) in the discretized domain
N	number of finite elements
N_i	shape function associated with node i

N_i^e	shape function for element e associated with node i
N_i^g	global shape function associated with node i
N_x, N_y, N_{xy}	membrane stress resultants
$p(x)$	distributed lateral loading on beam
$P_1, P_2, \ldots$ etc.	points
q	distributed pressure loading
q^ϵ	heat flux in homogenization
$\bar{q}$	volumetric average of flux
Q_x, Q_y	shear forces
r_{min}	minimum allowed size of elements of structure
R_r	rejection ratio
R_e	initial rejection ratio
s	bone remodelling parameter to define lazy zone
s_i	typical design variable
s_j	material parameters of element j
s_k	slack variable
s_i^l	lower bounds of design variable
s_i^u	upper bounds of design variable
s_i^0	parameter of the design model of element i
S	flexural or bending rigidity, $= EI$
t	thickness
t_0	thickness of main plate
t_1	thickness of reinforcement layer
t_i	thickness at node i
T^ϵ	temperature in homogenization
T	variation of temperature with respect to a given reference
u, v	global displacements
u, v, w	displacements of plate
u^0, v^0, w^0	mid-plane displacements of plate
u_i, v_i	nodal degrees of freedom associated with node i
u_e	effective strain energy density
u_n	homeostatic strain energy density
$\tilde{u}_i^1$	constant function of integration
U	the strain energy of the assemblage
v_i	volume (area) of element i
V	volume
$V_{initial}$	initial volume of the structure
W	weight of the structure
$\bar{w}_e$	filtered strain energy of element e
w_i	a function of design variables (e.g. displacement, stress, etc.)
$w(\mathbf{s})$	strain energy of element i
x_i, y_i	typical coordinates of node i
x_i^ℓ, x_i^u	lower and upper bounds of the design variable x_i

x	global cartesian coordinate
y	global cartesian coordinate
$y(x)$	vertical deflection of beam
$\bar{y}(x)$	Lagrange multiplier
Y_1, Y_2, Y_3	dimensions of the base cell of periodicity
$z(x)$	height of cross section of a beam
z^o	lower limit of z

Vectors

$\mathbf{b}_j$	position vector of control vertices
$\mathbf{d}$	vector of unknown displacements
$\mathbf{d}_i^e$	displacement associated with element e and node i
$\mathbf{d}_1$, $\mathbf{d}_2$, $\mathbf{d}_3$	column vectors of constitutive matrix
$\mathbf{f}$	vector of structural loads
	body forces
$\mathbf{f}_k^\star$	pseudo-load vector
$\mathbf{f}_i^e$	force vector for element e associated with node i
$\mathbf{g}$	vector of distributed loadings
$n_j, (j = 1, 2, 3)$	unit normal vector
$\mathbf{p}$	applied traction forces inside the hole of the base cell
	external loading
$\mathbf{p}_i$	typical point of contour
$\mathbf{q}$	a general force vector
$\mathbf{q}_i$	typical point of Bezier spline
$\mathbf{s}$	vector of design variable
$\mathbf{S}$	vector of parameters of analysis elements
$\mathbf{S}^0$	vector of parameters of design elements
$\mathbf{t}$	tractions
$\bar{\mathbf{u}}$	vector of prescribed displacements
$u^i(x, y)$, $(i = 1, 2, \cdots)$	displacements which are periodic on y
$\mathbf{x}$	design variables $\mathbf{x} = [x_1, x_2, \cdots, x_n]^T$
	position vector of a typical point
$\mathbf{Y} = [Y_1,\ Y_2,\ Y_3]^T$	vector of periodicity
$\mathbf{y} = \mathbf{x}/\epsilon$	position vector in local or microscopic coordinate

Matrices

$\mathbf{a}(\theta^e)$	rotation matrix, $\mathbf{a}(\theta^e) = \begin{bmatrix} \cos\theta^e & -\sin\theta^e \\ \sin\theta^e & \cos\theta^e \end{bmatrix}$
$\mathbf{B}$	strain-displacement matrix
$\mathbf{C}$	inverse of $\mathbf{D}$
	constitutive matrix of plate
$\mathbf{C}_s$	matrix of transverse shear coefficients
$\mathbf{D}$	constitutive matrix

$\mathbf{D}^H$	homogenized constitutive matrix
$\mathbf{D}_m$	matrix of membrane rigidities
$\mathbf{D}_f$	matrix of flexural rigidities
$\mathbf{D}_s$	matrix of shear rigidities
$\mathbf{D}_m^H$	matrix of homogenized membrane rigidities
$\mathbf{D}_f^H$	matrix of homogenized flexural rigidities
$\mathbf{D}_s^H$	matrix of homogenized transverse shear rigidities
$\mathbf{D}^0$	constitutive matrix for homogenous solid
$\mathbf{H}$	impulse response matrix
$\mathbf{J}$	Jacobian
$\mathbf{K}$	symmetric, banded stiffness matrix
$\mathbf{k}_{ij}^e$	submatrix of stiffness matrix of element e associated with nodes i and j
$\mathbf{K}_{ij}^e$	stiffness matrix associated with element e and nodes i and j
$\mathbf{K}_{mij}^e$	membrane stiffness matrix for element e and nodes i and j
$\mathbf{K}_{bij}^e$	bending stiffness matrix of element e and nodes i and j
$\mathbf{K}_{sij}^e$	shear stiffness matrix of element e and nodes i and j
$\mathbf{K}_{pij}$	in-plane stiffness matrix linking node i and j
$\mathbf{L}$	matrix of linear differential operators linking matrix between analysis and design element parameters
$\mathbf{N}$	matrix of arbitrary integers $= \begin{bmatrix} n_1 & 0 & 0 \\ 0 & n_2 & 0 \\ 0 & 0 & n_3 \end{bmatrix}$
$\mathbf{N}_i^e$	shape function matrix for element e assciated with node i
$\mathbf{N}_i^g$	global shape function matrix associated with node i
$\mathbf{N}^g$	global shape function matrix
$\mathbf{R}(\theta^e)$	rotation matrix for element e
$\mathbf{T}$	transformation matrix

Greek symbols: scalars

α	parameter of bone remodelling simulation thermal expansion coefficient
β_1, β_2	normal rotations
β	incremental growth rate
γ	parameter of bone remodelling simulation relative density of solid layer in ranked layered material cell
$\gamma(y)^\epsilon$	weight per unit volume of material
$\overline{\gamma}$	volumetric average of γ inside the base cell, $\overline{\gamma} = \frac{1}{Y}\int_Y \gamma(y)dy$
Γ	boundary Lagrangian multiplier
Γ_d	boundary on which displacements are prescribed
Γ_t	boundary on which tractions are prescribed

δ	acceptable tolerance for the volume constraint
δ_i	mesh density
$\Delta\pi$	variations of bone parameter π
Δs_k	perturbation of design variable s_k
ϵ	small parameter indicating the characteristic inhomogeneity dimension
	as a superscript indicates dependency to the cell of periodicity
ϵ_{ij}	strains
ϵ_{ij}^0	initial strains
ζ	move limit
η	damping factor
	a dummy variable of integration
	local refining indicator
η_{min}, η_{max}	lower and upper bounds of the refining indicator
θ	orientation of microscale perforations in homogenization
θ_{xi}, θ_{yi}	global rotational degrees of freedom at node i
θ_i	rotational degree of freedom at node i
κ	shear modification factor taken as $\frac{5}{6}$ for isotropic materials
λ_j, λ_{a0}^e, λ_{a1}^e	Lagrangian multiplier
λ_{b0}^e, λ_{b1}^e	Lagrangian multiplier
$\Lambda_{min}^0, \Lambda_{max}^0$	starting Lagrange multipliers in bisection method
Λ^m	average of Λ_{min} and Λ_{max} at iteration m
Λ_k	Lagrangian multiplier
μ	rank-2 material parameter
	exponent used in artificial material model
ν	Poisson's ratio
ν_k	Lagrangian multiplier
ξ	isoparametric element natural coordinate
$\xi(\mathbf{x})$	continuous approximation to indicator function $\chi(\mathbf{x})$
ξ_1, ξ_2	independent artificial density functions
π	a parameter that defines the bone properties
Π	total potential energy
Π^e	potential energy contribution from element e
ρ	density
ρ_m	density of mineralized tissue
ρ_a	apparent density
ρ_s	density of solid part
ρ^0	density of homogeneous solid
ρ_{lim}	limiting value of material density

$\rho(k,\ell)$	the density of pixel located in the kth row and ℓth column
$\rho^*(i,j)$	filtered density of pixel
σ	stress component
σ^ϵ	axial stress in homogenization problem of one dimensional elasticity
σ_{ij}	second order stress tensor
σ_{all}	allowable stress
σ_{eff}	effective stress
σ_{avg}	average stress
σ_{bas}	reference stress
σ_{von}	von Mises stress
σ_{eq}	equivalent stress
$\sigma^i(x,y)$, $(i=1,2,\cdots)$	stresses which are periodic on y and the length of period is Y
τ_{ij}	average 'residual' stresses within the cell due to the tractions
ϕ_j	blending function
Φ	criterion function
$\Phi(x)$	an oscillating function
$\Phi^0(\mathbf{x},\mathbf{y})$, $\Phi^1(\mathbf{x},\mathbf{y})$, $\cdots$	functions which are smooth with respect to $\mathbf{x}$ and periodic in $\mathbf{y}$
$\chi(y)$	an initial function
χ^{kl}	Y-periodic function
$\chi(\mathbf{x})$	indicator function
Ω	domain of problem
Ω_ℓ	volume of element ℓ
Ω^ϵ	solid part of cell in homogenization
Ω_s	solid domain
$\bar{\Omega}_s$	given volume for solid

Greek symbols: vectors and matrices

γ	acceleration vector
$\boldsymbol{\theta}$	displacement function
τ_{xy}	shear stress
$\boldsymbol{v}$	kinematically admissible virtual displacement field
$\boldsymbol{\Phi}$	displacement function
$\hat{\boldsymbol{\Phi}}$	discretized displacement function
$\boldsymbol{\chi}$	microscopic characteristic displacement field, Y-periodic solution of $\mathbf{K}\,\boldsymbol{\chi}=\mathbf{f}$
Ψ	a general periodic function
$\boldsymbol{\Psi}$	displacement function
$\hat{\boldsymbol{\Psi}}$	discretized displacement function
ϵ_m	vector of membrane strains
ϵ_f	vector of bending strains or curvatures

$\epsilon_m, \epsilon_f, \epsilon_s$	membrane, bending and transverse shear strains
σ	stress resultant vector
$\sigma_m, \sigma_f, \sigma_s$	membrane, bending and shear stress resultant vectors

Special symbols:

$\overset{m}{\underset{e=1}{\mathbf{A}}}$	finite element assembly operator
$\overline{A}$	arithmetic average of two real numbers
$\underline{A}$	harmonic average of two real numbers
$\mathcal{A}^\epsilon$	elliptical operator, $\mathcal{A}^\epsilon = \frac{\partial}{\partial x_i}\left(a_{ij}(\mathbf{y})\frac{\partial}{\partial x_j}\right)$
$\mathcal{A}^1, \mathcal{A}^2, \mathcal{A}^3$	differential operators
$\mathbb{D}$	set of admissible shapes
$f(.)$	function of
H^1	Sobolev space
L_2	Hilbert space
$\mathbb{R}^n$	n dimensional space
$\mathbf{V}$	space of admissible displacements
$\mathbf{V}_\Omega$	space of admissible displacements for vectors defined in Ω
$\mathbf{V}_{¥}$	space of admissible displacements for vectors defined on the solid part of cell
$\mathbf{V}_{\Omega\times ¥}$	space of admissible displacements
$\mathbf{Y}$	base cell of periodicity
¥	solid part of the cellular base cell of periodicity
δ_{ij}	Kronecker delta symbol
∇	gradient operator
$\partial\ell$	partial differential of ℓ
$\langle f\rangle$	volumetric average of function f over the period Y
$\lVert.\rVert$	norm (e.g. Euclidean, absolute, etc)
$\in$	belongs to
$\sum$	summation
$\bigcup$	union of sets
$\cap$	intersection of sets
$\exists$	exists
$\forall$	for all
$'$	implies differentiation with respect to x
$\emptyset$	empty set

1

INTRODUCTION

The objective of this chapter is to give an overview of structural topology optimization. The mathematical description of a general optimization problem is explained. The types of structural optimization are briefly reviewed and some aspects of the structural topology optimization are introduced. Finally, the layout of the book is outlined.

1.1 Overview

Since the resources available to human beings in nature are limited, it is very important to use these resources as efficiently as possible. In this respect optimization, which can be simply defined as the process of searching for the 'best', serves as a valuable tool. The main purpose of engineering design is to find the best possible solution to a specific problem, optimization is therefore at the very heart of engineering.

Engineering design has traditionally been a process of trial and error. It starts from a primary design which is based on the creativity, past experience, intuition and knowledge of the designer. This is followed by an analysis to evaluate the performance of the design. Based upon the information provided by this analysis, a new design is developed. The steps: design - analysis - new design are repeated until no further improvement is possible.

The advances in computer technology and the relatively inexpensive computational power has provided the opportunity to revolutionize the traditional design process. This has already occurred in some engineering fields such as structural engineering where 'optimization' is introduced into the design process instead of the intuitive heuristic repeated trials of the traditional method [1].

The task of structural optimization is to find the 'best possible' or 'optimal' structure which meets all the multidisciplinary requirements such as those imposed by functionality and manufacturing conditions. A structural optimization problem may be defined as the rational establishment of a structural design that is the best of all possible designs within a prescribed objective and a given set of limitations [2].

In structural analysis, the geometry and dimensions are known and the purpose of analysis is to determine the behaviour and response (e.g. stresses, displacements, frequencies, etc) of the structure. In contrast, in structural optimization certain dimensions and geometrical or material properties of the structure are unknowns. The aim is to find the minimum (or maximum) of a predefined objective function subject to some limitations and constraints. The objective function can be the weight or stiffness of the structure, cost of material, storage capacity and so forth. The constraints could be *geometrical* such as restrictions on the width or height of the structure or *behavioural* such as restrictions on stresses, displacements and natural frequencies.

Various optimum design techniques may be classified as *intuitive*, methods based on *optimality criteria* and methods based on *mathematical programming*. Examples of the intuitive methods are the principles of fully stressed design and simultaneous failure design which were used to a great extent throughout this century particularly in design of air frame structures. The optimality criteria methods consist of two steps: (a) the identification of a set of necessary conditions that must be satisfied at the optimum design and (b) the development of an iterative redesign procedure to reach that optimum. In the mathematical programming methods, structural design is treated as a problem of mathematical extremization of a cost (or objective) function in a multidimensional design variable space, constrained by some behavioural and/or geometrical functions [3].

1.2 Mathematical description of optimization problem

An optimization problem seeks the minimum (or maximum) value of a function $f(\mathbf{x})$ and the related variable vector $\mathbf{x} = (x_1, \cdots, x_n) \in \mathbb{R}^n$ which yields it. Here, f is called the *objective (or cost) function.* The quantities $x_i,\ i = 1, \cdots, n$ are the variables of the optimization problem, usually called *design variables* and n is the number of design variables. Note that any vector $\mathbf{x}$ in the n-dimensional design space $\mathbb{R}^n$ represents a single design.

It may happen that the design variables are not totally free and they have to satisfy some extra conditions or constraints. Most of the optimization problems confronted in practice are of this kind. These constraints may have an equality form. This indicates that the design variables are not independent and, in fact, the real design space is a sub-space of $\mathbb{R}^n$, where the dimension of this space will be n minus the number of the independent equality constraints. The strict inequality constraints impose some limitations on the values of the design variables $x_i \in \mathbb{R}$ and comprise a subset of the design space.

The optimization problem in its most general form may be expressed as

$$\begin{array}{lll} \text{Minimize} & f(\mathbf{x}) & \\ \text{such that} & h_j(\mathbf{x}) = 0, & j = 1, 2, \cdots, n_h \\ & g_k(\mathbf{x}) \leq 0, & k = 1, 2, \cdots, n_g \\ & x_i^\ell \leq x_i \leq x_i^u, & i = 1, 2, \cdots, n \end{array} \qquad (1.1)$$

where n_h, n_g and n are the number of equality constraints, inequality constraints and design variables, respectively. The values x_i^ℓ and x_i^u are the lower and upper bounds on a typical design variable x_i.

The set of design variables which satisfy all the constraints comprise the *feasible domain.* The *infeasible domain* is the collection of all design points that violate at least one of the constraints.

If the objective function and both equality and inequality constraints are linear functions of the design variables, then the problem is a *linear optimization problem.* In a *non-linear optimization problem,* either the objective function or at least one of the constraints is a non-linear function of the design variables. Structural optimization problems are usually non-linear optimization problems.

1.3 Types of structural optimization

For any structural design process three stages could be considered. The first, a *conceptual* stage, where the structural system and form is chosen. This stage involves applying experience and judgement in a qualitative manner and does not have an algorithmic basis and is therefore difficult to computerize. The second, a *preliminary* design stage, where the shape of the structure and the defining geometry is decided. The final stage, is a *detailed* design [4].

The early research on structural optimization focussed on *sizing optimization* problems which involves, for example, finding the optimum cross sectional properties of members of a truss or frame structure or the thickness optimization of a plate structure. In these problems the domain is fixed and does not change during the optimization process. Sizing optimization may be considered as the implementation of optimization at the detailed design stage.

As a further development the problem of finding optimal boundaries of a structure was considered. Examples of this type of problem include: finding the boundaries of a structure modelled using plane stress assumptions or finding the location of joints of a skeletal structure or finding the optimal values for parameters which define the middle surface of a shell structure. In the literature this class of structural optimization problems is referred to as *shape optimization*. In these problems the shape of the domain is not fixed but the topology[†] is. This may be seen as the implementation of the optimization techniques at the preliminary design stage.

The sizing and shape optimization methods may lead to sub-optimal results as they suffer from the problem of not necessarily having an optimum starting topology. To overcome this deficiency *topology optimization* must be considered. Typically in topology optimization of 2D and 3D continua the aim is to determine features such as the number of holes and their location. The problem of finding the optimum configuration and spatial sequence of members and joints of a skeletal structure also lies in this category. The ideal is to find a method to simultaneously optimize the geometry (i.e. size and shape) and topology of the structure. This is sometimes called *layout optimization* [5]. This class of optimization problem can be thought of as optimization at the conceptual stage.

[†] Topology is a branch of mathematics, sometimes given the nickname 'rubber - sheet' geometry. Two geometrical figures that can be transformed from one to another through 'continuous transformations' are said to have the same topology or are topologically equivalent. By 'continuous transformation' we mean pulling, stretching, twisting, bending or squashing without tearing or 'gluing' points together [6].

1.4 Aspects of topology optimization

Topological structural optimization has the complex features of both size and shape optimization problems. Conventional shape optimization by the 'boundary variation method' normally requires several re-meshings in the optimization process and results in final designs that are topologically equivalent to the initial design. Trying to change the topology as well as the shape during the scheme will increase the complexity of the problem and will make it a very difficult task. Because of these complexities, this class of problems is regarded as one of the most challenging in structural mechanics [7].

Methods for obtaining optimal topologies vary from rigorous mathematically based methods (e.g. the homogenization method) to more engineering-intuitive methods (e.g. hard kill/soft kill method). In this book the focus will be on the former approaches.

Usually the result of structural topology optimization for 2D and 3D continua is a contour plot of material density from which an optimal topology must be discerned either automatically using some kind of image processing or intuitively using engineering judgement or by a combination of both.

In an attempt to solve the topology optimization problem Bendsøe and Kikuchi [8] suggested the *homogenization method* which has since attracted the attention of many researchers. In this method the optimal topology is accompanied by a rough optimal shape and size and consequently, it is sometimes called *generalized shape optimization.*

Consideration of the following factors provided the inspiration for the homogenization method:

- Generalized shape optimization is inherently a point-wise 'material/no material' problem.
- Implementation of this 'on-off' approach to an optimization problem requires the use of discrete optimization algorithms and such approaches can be unstable.
- The experience of previous researchers has proved that in many cases the optimum result contains regions with infinitesimal cavities or ribs [9, 10].
- By introducing a microstructure to the material model, Kohn and Strang [11] could obtain a well-posed, relaxed formulation for the two dimensional heat conduction problem where the solution was no longer mesh-dependent.
- The mathematical limit theory of homogenization based on asymptotic expansions has been well developed [12].
- Parallel to these advances, optimality criteria methods with the ability to handle optimization problems with a large number of design variables have reached a mature level [5].

In the homogenization method, microstructures are introduced to the material of the finite elements of the discretized domain and the parameters of these microstructures are treated as the design variables of the optimization problem. For this purpose a few material models have been developed. Some of the most common ones are introduced in Chapter 3. In the homogenization method the shape of the structure is represented without geometrical segments (i.e. lines, circular arcs, elliptic arcs and splines) and it is not necessary to change the finite element mesh which has been defined for the analysis during the optimization process.

The outcome of topological structural optimization using homogenization theory and optimality criteria methods, which is usually presented in the form of a material density plot, is affected by several factors such as the material model, the finite element mesh, the type of elements, the resizing scheme for updating the design variables etc. One should note that no proof of the convexity of the topology optimization problem is yet available.

1.5 Layout of the book

This book is divided into three parts. Part I comprises Chapters 2 and 3. The homogenization theory for media with periodic structure is dealt with in Chapter 2. The main objective of this chapter is to derive the required homogenization equations which must be solved. The main concepts of the theory: periodicity and asymptotic expansion, are defined and the homogenization problem for various cases such as one dimensional elasticity, three dimensional heat conduction, general boundary value problems in strong form and two dimensional elasticity problems in weak form are discussed.

In Chapter 3 different material models are introduced which are suitable for use in structural topological optimization such as the unit square cell with a rectangular void, the rank layered material and the artificial material. The analytical solution of the homogenization equation is considered for rank laminate composites comprising rank-1, rank-2 and bi-material rank-2 representations. The finite element formulation for the numerical solution of the equations is derived and a direct method to deal with the issue of boundary conditions for bi-symmetric cells of periodicity is proposed and some examples are provided. The results obtained for the material model with rectangular voids are also given together with the associated bi-cubic and bi-quartic polynomials resulting from a conventional least squares data fitting. Thus, at this stage the relationship between the components of the homogenized elastic constitutive matrix and the hole size dimensions is available in polynomial form.

Part II is devoted to topology optimization of plane and plate structures using the homogenization and optimality criteria methods. Chapters 4, 5 and 6 form Part II of the book. The optimality criteria method is the concern of Chapter 4. The Kuhn-Tucker optimality conditions are explained and via examples, the analytical optimality criteria methods are briefly introduced. A mathematical model for the topological structural optimization problem is constructed with the objective of total potential energy to be maximized. The optimality criteria are also derived. A modified resizing scheme is suggested. The issue of different schemes for optimal orientation is discussed. The algorithm of the developed system **PLATO** (PLAne Topology Optimization), which is provided with this book on CD-ROM, is explained and some examples are provided.

A systematic study of the parameters which have an influence on the outcome of the topological optimization process is carried out in Chapter 5 and several problems using plane stress assumptions are provided.

The extension of the homogenization and penalty methods to layout and reinforcement optimization of plate structures in a format which may easily be extended to shell problems is the concern of Chapter 6. Selection of the plate base cell models is discussed and the underlying Mindlin-Reissner plate theory is briefly reviewed. The homogenization of the plate microstructure model is considered and a modification to take into account the transverse shear resistance of the homogenized reinforcement material is suggested. A hierarchical version of the heterosis elements based on Mindlin-Reissner plate theory to be used in FE analysis, is explained. Finally, several illustrative examples are presented.

Part III consists of Chapters 7 and 8 which present some alternative methods and the concept of 'integrated structural optimization'. Alternative approaches to the structural optimization problem are considered in Chapter 7. The 'simulation of functional adaptation of bone mineralization' is presented with remodeling schemes based on effective strain energy density and effective stress level and the algorithm used in structural topology optimization is explained. The 'evolutionary fully stressed method' is also discussed and a few examples are provided.

In Chapter 8, the idea of an 'integrated structural optimization system' is dealt with and its three modules are explained. In the topology optimization module, the 'ground structure' and 'bubble' methods are briefly explained. In the 'image processing' module a method for the elimination of mesh dependency and checkerboard problems by use of image processing techniques is briefly introduced. In the 'shape optimization' module, the well known 'boundary variation method' is briefly reviewed and as an alternative, the 'adaptive growth method' is explained and some examples are provided. Finally, the 'adaptive' approach to integrated structural topology and shape optimization is briefly presented.

References

[1] Frangopol D.M. and Cheng F.Y., Advances in structural optimization, Tech. rep. (1996)

[2] Olhoff N. and Taylor J.E., On structural optimization, *J. of Applied Mechanics*, **50**, 1139–1151 (1983)

[3] Chibani L., Optimum design of structures, in *Lecture notes in engineering*, edited by Brebbia C.A. and Orszag S.A., vol. 50. Springer-Verlag, Berlin (1989)

[4] Grierson D.E., Conceptual design using emerging computing techniques, Tech. rep., NATO Advanced Research Workshop, Nafplio, Greece (1994)

[5] Rozvany G.I.N (ed.), *Shape and layout optimization of structural systems and optimality criteria methods*. Springer-Verlag, CISM, Udine (1992)

[6] Firby P.A. and Gardiner C.F., *Surface topology, 2nd edn.* Ellis Horwood, Chichester (1991)

[7] Rozvany G.I.N., Bendsøe M.P. and Kirsh U., Layout optimization of structures, *Appl. Mech. Rev.*, **48**(2), 41–119 (1995)

[8] Bendsøe M.P. and Kikuchi N., Generating optimal topologies in structural design using homogenization method, *Comp. Meth. Appl. Mech. Eng.*, **71**, 197–224 (1988)

[9] Lurie K.A. and Chirkaev A.V., G-Clousure of some particular sets of admissible material characteristics for the problem of bending of thin plates, *J. Optim. Theory Appl.*, **42**, 305–315 (1984)

[10] Rozvany G.I.N., Olhoff N., Bendsøe M.P., Ong T.G., Sandler R. and Szeto W.T., Least-weight design of perforated elastic plates, Parts I and II, *Int. J. Solids Struct.*, **23**, 521–536, 537–550 (1987)

[11] Kohn R.V. and Strang G., Optimal design and relaxation of variational problems, *Comm. Pure Appl. Math.*, **39**, 113–137 (Part I), 139–182 (Part II), 353–377 (Part III) (1986)

[12] Sanchez-Palencia E., Non-homogenous media and vibration theory, *Lecture Notes in Physics*, **127** (1980)

Part I

Homogenization

2

HOMOGENIZATION THEORY FOR MEDIA WITH PERIODIC STRUCTURE

In this chapter an overview of the theory of homogenization for composites with regular structure is presented. Periodicity and asymptotic expansion are defined and an application of homogenization to the simple case of a one dimensional elasticity problem is given. Derivation of the basic formulas for the general case of a boundary value problem in strong form is discussed. Finally, the homogenization equations for the elasticity problems in weak form for perforated media are derived.

2.1 Introduction

Advances in technology in recent years have been paralleled by the increased use of composite materials in industry. Since materials have different properties, it seems sensible to make use of the good properties of each single ingredient by using them in a proper combination. For example, a simple mixture of clay, sand and straw produced a composite building material which was used by the oldest known civilizations. The further development of non-metallic materials, composites has attracted the attention of scientists and engineers in various fields, for example, aerospace, transportation, and other branches of civil and mechanical engineering. Apart from the considerably low ratio of weight to strength, some composites benefit from other desirable properties such as corrosion and thermal resistance, toughness and lower cost. Usually composite materials comprise a *matrix* which could be metal, polymeric (like plastics) or ceramic, and a *reinforcement* or *inclusion* which could be particles or fibres of steel, aluminum, silicon etc.

Composite materials may be defined as a man-made material with different dissimilar constituents which occupy different regions with distinct interfaces between them [1]. The properties of a composite are different from its individual constituents. A cellular body can be considered as a simple case of a composite, comprising solids and voids. This is the case for the material model used in structural topology optimization.

In this chapter composites with a regular or nearly regular structure are considered. Having sufficiently regular heterogenities enables us to assume a periodic structure for the composite. It should be emphasised that compared with the dimensions of the body the size of these non-homogeneities should be very small. Because of this, these types of material are sometimes called *composites with periodic micro-structures.*

Even with the help of high-speed modern computers, the analysis of the boundary value problems consisting of such media with a large number of heterogenities, is extremely difficult. A natural way to overcome this difficulty is to replace the composite with a kind of equivalent material model. This procedure is usually called *homogenization.* One way of finding the properties of such composites is by doing experimental tests. It is quite evident that because of the volume and cost of the required tests for all possible reinforcement types, experimental measurements are often impracticable.

The mathematical theory of homogenization, which has developed since the 1970's is used as an alternative approach to find the effective properties of the equivalent homogenized material [2-4]. This theory can be applied in many areas of physics and engineering having finely heterogeneous continuous media like heat transfer or fluid flow in porous media or, for example, electromagnetism in composites. In fact the basic assumption of continuous media in mechanics and physics can be thought of as sort of homogenization, as the materials are composed of atoms or molecules.

From a mathematical point of view the theory of homogenization is a limit theory which uses the asymptotic expansion and the assumption of periodicity to substitute the differential equations with rapidly oscillating coefficients with differential equations whose coefficients are constant or slowly varying in such a way that the solutions are close to the initial equations [5].

This method makes it possible to predict both the overall and local properties of processes in composites. In the first step the appropriate local problem on the unit cell of the material is solved and the effective material properties are obtained. In the second step the boundary value problem for the homogenized material is solved.

2.2 Periodicity and asymptotic expansion

A heterogeneous medium is said to have a regular periodicity if the functions denoting some physical quantity of the medium - either geometrical or some other characteristics - have the following property

$$\mathcal{F}(\mathbf{x}+\mathbf{NY}) = \mathcal{F}(\mathbf{x}). \tag{2.1}$$

$\mathbf{x} = [x_1, x_2, x_3]^T$ is the position vector of the point, $\mathbf{N}$ is a 3×3 diagonal matrix

$$\mathbf{N} = \begin{bmatrix} n_1 & 0 & 0 \\ 0 & n_2 & 0 \\ 0 & 0 & n_3 \end{bmatrix},$$

where n_1,n_2 and n_3 are arbitrary integer numbers and $\mathbf{Y} = [Y_1, Y_2, Y_3]^T$ is a constant vector which determines the period of the structure and $\mathcal{F}$ can be a scalar or vectorial or even tensorial function of the position vector $\mathbf{x}$. For example in a composite tissued by a periodically repeating cell $\mathbf{Y}$, the mechanical behavior is described by the constitutional relations of the form

$$\sigma_{ij} = c_{ijkl}e_{kl}$$

and the tensor c_{ijkl} is a periodic function of the spatial coordinate $\mathbf{x}$, so that

$$c_{ijkl}(\mathbf{x}+\mathbf{NY}) = c_{ijkl}(\mathbf{x}) \tag{2.2}$$

or

$$c_{ijkl}(x_1 + n_1Y_1, x_2 + n_2Y_2, x_3 + n_3Y_3) = c_{ijkl}(x_1, x_2, x_3).$$

$c_{ijkl}(\mathbf{x})$ is called $\mathbf{Y}$-periodic (see Figure 2.1). Note that σ_{ij} and e_{kl} are respectively the stress and strain tensors.

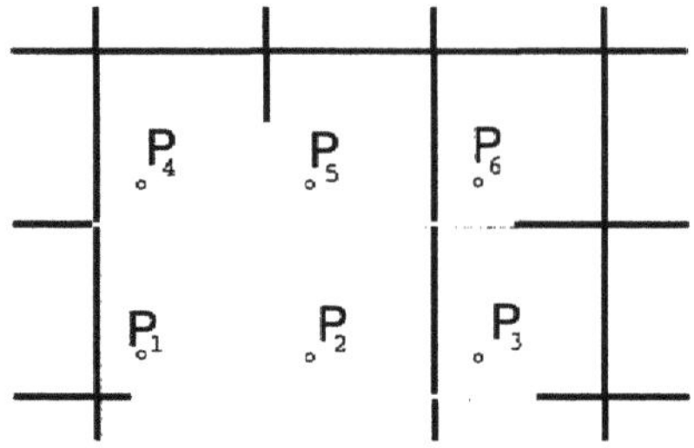

Figure 2.1 *Periodicity requires that the functions have equal values at points $P_1, P_2, \cdots, P_6$.*

In the theory of homogenization the period $\mathbf{Y}$ compared to the dimensions of the overall domain is assumed to be very small. Hence the characteristic functions of these highly heterogeneous media will rapidly vary within a very small neighbourhood of a point $\mathbf{x}$. This fact inspires the consideration of two different scales of dependencies for all quantities: one on the *macroscopic* or *global* level $\mathbf{x}$ which indicates *slow* variations and the other on the *microscopic* or *local* level $\mathbf{y}$ which describes *rapid* oscillations.

The ratio of the real length of a unit vector in the microscopic coordinates to the real length of a unit vector in the macroscopic coordinates, is a small parameter ϵ; so $\epsilon\mathbf{y} = \mathbf{x}$ or $\mathbf{y} = \mathbf{x}/\epsilon$. Consequently, if g is a general function then we can say $g = g(\mathbf{x}, \mathbf{x}/\epsilon) = g(\mathbf{x}, \mathbf{y})$. To illustrate the technique let us assume that $\Phi(x)$ is a physical quantity of a strongly heterogeneous medium. Thus $\Phi(x)$ will have oscillations. See Figure 2.2 .

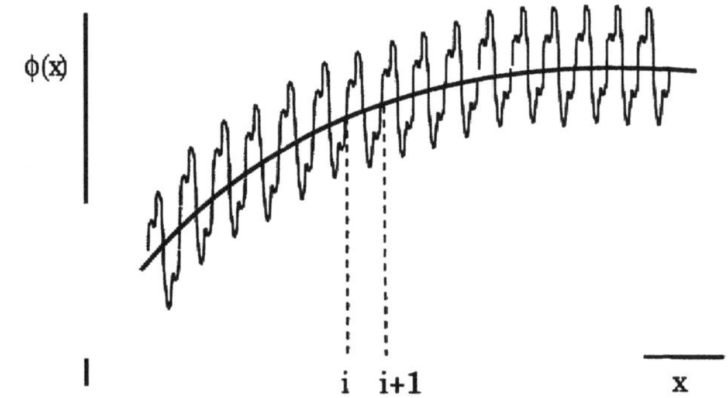

Figure 2.2 *A highly oscillating function*

To study these oscillations using this *double-scale* expansion, the space can be enlarged as indicated in Figure 2.3 .

The small parameter ϵ also provides an indication of the proportion between the dimensions of the base cells of a composite and the whole domain known as the *characteristic* inhomogeneity dimension. As a hypothetical example, ϵ for the skin cells of the human body is larger than ϵ for the atoms of which it is

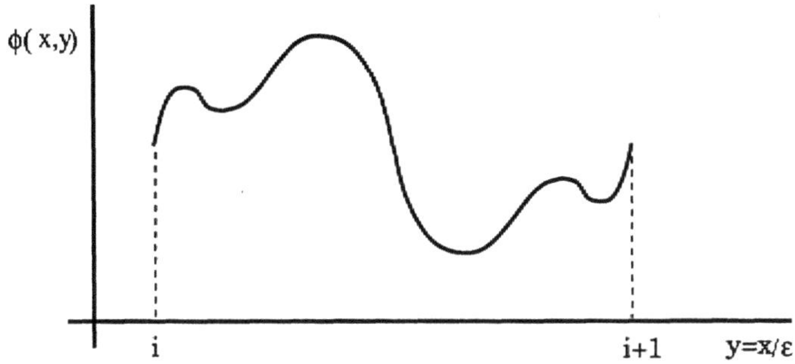

Figure 2.3 *One of the oscillations in the expanded scale*

composed. The quantity $1/\epsilon$ can be thought of as a magnification factor which enlarges the dimensions of a base cell to be comparable with the dimensions of the material [6-8]. See Figure 2.4 .

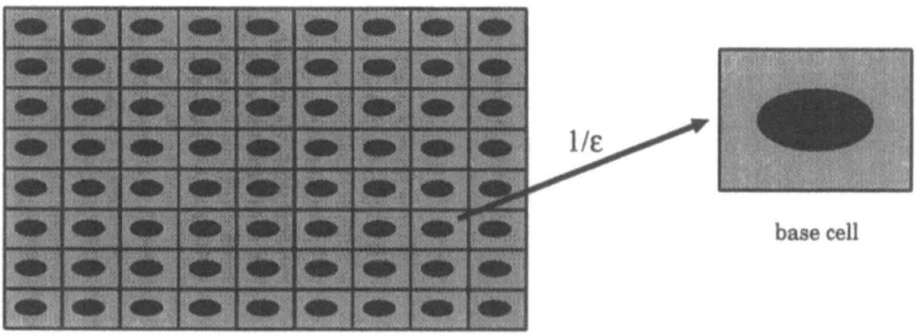

Figure 2.4 *Characteristic dimension of inhomogeneity and scale enlargement*

In the double-scale technique, the partial differential equations of the problem have coefficients of the form $a(\mathbf{x}/\epsilon)$ or $a(\mathbf{y})$ where $a(\mathbf{y})$ is a periodic function of its arguments. The corresponding boundary value problem may be treated by asymptotically expanding the solution in powers of the small parameter ϵ. This technique has already proved to be useful in the analysis of slightly perturbed periodic processes in the theory of vibrations. The same principle is extendible to processes occurring in composite materials with a regular structure.

If we assign a coordinate system $\mathbf{x} = (x_1, x_2, x_3)$ in $\mathbb{R}^3$ space to define the domain of the composite material problem Ω, then assuming periodicity, the domain can be regarded as a collection of parallelepiped cells of identical dimensions $\epsilon Y_1, \epsilon Y_2, \epsilon Y_3$, where Y_1, Y_2 and Y_3 are the sides of the base cell $\mathbf{Y}$ in a local (microscopic) coordinate system $\mathbf{y} = (y_1, y_2, y_3) = \mathbf{x}/\epsilon$. So for a fixed $\mathbf{x}$ in the macroscopic level, any dependency on $\mathbf{y}$ can be considered $\mathbf{Y}$-periodic. Moreover it is assumed that the form and composition of the base cell varies in a smooth way with the macroscopic variable $\mathbf{x}$. This means that for different points the structure of the composite may vary, but if one looks through a microscope at a point at $\mathbf{x}$, a periodic pattern can be found.

Functions determining the behavior of the composite can be expanded as

$$\Phi^{\epsilon}(\mathbf{x}) = \Phi^{0}(\mathbf{x},\mathbf{y}) + \epsilon\Phi^{1}(\mathbf{x},\mathbf{y}) + \epsilon^{2}\Phi^{2}(\mathbf{x},\mathbf{y}) + \cdots$$

where $\epsilon \to 0$ and functions $\Phi^{0}(\mathbf{x},\mathbf{y})$, $\Phi^{1}(\mathbf{x},\mathbf{y})$, $\cdots$ are smooth with respect to $\mathbf{x}$ and $\mathbf{Y}$-periodic in $\mathbf{y}$ which means that they take equal values on the opposite sides of the parallelepiped base cell.

2.3 One dimensional elasticity problem

To clarify the homogenization method the simple case of calculation of deformation of an inhomogeneous bar in the longitudinal direction is considered. Here we attempt to derive the modulus of elasticity without recourse to advanced mathematics.

According to the assumptions of the theory, the medium has a periodic composite microstructure. (See Figure 2.5).

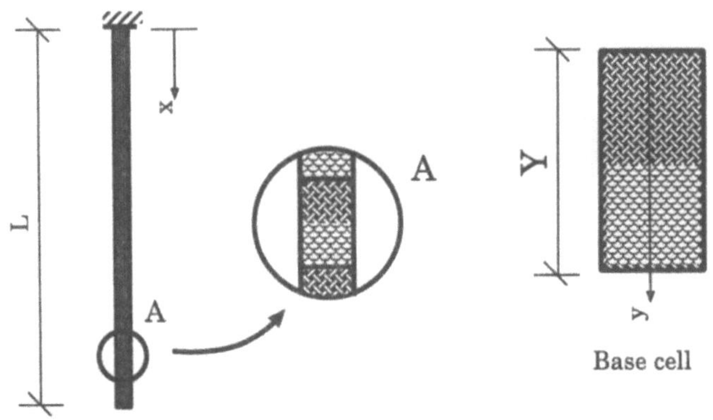

Figure 2.5 Deformation of a composite bar

The governing equations, in the form of Hooke's law of linear elasticity and the Cauchy's first law of motion (or equilibrium equation) , are

$$\sigma^{\epsilon} = E^{\epsilon}\frac{\partial u^{\epsilon}}{\partial x}, \tag{2.3}$$

$$\frac{\partial \sigma^{\epsilon}}{\partial x} + \gamma^{\epsilon} = 0. \tag{2.4}$$

The dependency of the quantities to the size of the unit cell of inhomogeneity is indicated by the superscript ϵ. σ^{ϵ} is the stress, u^{ϵ} is the displacement, $E^{\epsilon}(x)$

is the Young's modulus and γ^ϵ is the weight per unit volume of material. It is assumed that E^ϵ and γ^ϵ are macroscopically uniform along the domain and only vary inside each cell,

$$E^\epsilon(x, x/\epsilon) = E^\epsilon(x/\epsilon) = E(y) \tag{2.5}$$

and

$$\gamma^\epsilon(x, x/\epsilon) = \gamma^\epsilon(x/\epsilon) = \gamma(y). \tag{2.6}$$

Using the double scale asymptotic expansion

$$u^\epsilon(x) = u^0(x,y) + \epsilon u^1(x,y) + \epsilon^2 u^2(x,y) + \cdots \tag{2.7}$$

and

$$\sigma^\epsilon(x) = \sigma^0(x,y) + \epsilon\sigma^1(x,y) + \epsilon^2\sigma^2(x,y) + \cdots \tag{2.8}$$

where $u^i(x,y)$ and $\sigma^i(x,y)$, $(i = 1, 2, \cdots)$ are periodic on y and the length of period is Y. In due course the following facts will be referred to:

Fact (1). The derivative of a periodic function is also periodic with the same period.

Fact (2). The integral of the derivative of a function over the period is zero. (These facts can easily be verified by the definition of derivative and periodicity.)

Fact (3). If $\Phi = \Phi(x,y)$ and y depends on x, then

$$\frac{d\Phi}{dx} = \frac{\partial\Phi}{\partial x} + \frac{\partial\Phi}{\partial y}\frac{\partial y}{\partial x}.$$

In this case, as $y = x/\epsilon$ so

$$\frac{d\Phi}{dx} = \frac{\partial\Phi}{\partial x} + \frac{1}{\epsilon}\frac{\partial\Phi}{\partial y}.$$

Using the latter fact and substituting the series (2.7) and (2.8) into equations (2.3) and (2.4) we obtain,

$$\sigma^0 + \epsilon\sigma^1 + \epsilon^2\sigma^2 + .. = E(y)\left[\frac{\partial u^0}{\partial x} + \frac{1}{\epsilon}\frac{\partial u^0}{\partial y} + \epsilon\frac{\partial u^1}{\partial x} + \frac{\partial u^1}{\partial y} + \epsilon^2\frac{\partial u^1}{\partial x} + \epsilon\frac{\partial u^2}{\partial y} + ..\right] \tag{2.9}$$

and

$$\frac{\partial\sigma^0}{\partial x} + \frac{1}{\epsilon}\frac{\partial\sigma^0}{\partial y} + \epsilon\frac{\partial\sigma^1}{\partial x} + \frac{\partial\sigma^1}{\partial y} + \cdots + \gamma(y) = 0. \tag{2.10}$$

By equating the terms with the same power of ϵ, (2.9) yields

$$0 = E(y)(\frac{\partial u^0}{\partial y}), \tag{2.11}$$

$$\sigma^0 = E(y)(\frac{\partial u^0}{\partial x} + \frac{\partial u^1}{\partial y}), \tag{2.12}$$

$$\sigma^1 = E(y)(\frac{\partial u^1}{\partial x} + \frac{\partial u^2}{\partial y}), \tag{2.13}$$

and similarly from (2.10)

$$\frac{\partial \sigma^0}{\partial y} = 0, \tag{2.14}$$

$$\frac{\partial \sigma^0}{\partial x} + \frac{\partial \sigma^1}{\partial y} + \gamma(y) = 0. \tag{2.15}$$

From (2.11) and (2.14) it is concluded that the functions u^0 and σ^0 only depend on x (i.e. $u^0(x)$ and $\sigma^0(x)$). Bearing in mind that the relationship between $\sigma^0(x)$ and $u^0(x)$ is sought (because they are independent of the microscopic scale), (2.12) can be written as

$$\sigma^0(x) = E(y)\left[\frac{du^0(x)}{dx} + \frac{\partial u^1(x,y)}{\partial y}\right]. \tag{2.16}$$

Dividing by $E(y)$ and integrating both sides of (2.16) over the period Y and using fact (2) yields

$$\sigma^0(x) = \left(Y/\int_Y \frac{dy}{E(y)}\right)\frac{du^0(x)}{dx}. \tag{2.17}$$

Now by substituting the value of $\sigma^0(x)$ into (2.16), we obtain

$$\frac{\partial u^1(x,y)}{\partial y} = \left[Y/\left(E(y)\int_Y \frac{dy}{E(y)}\right) - 1\right]\frac{du^0(x)}{dx},$$

and by integrating this equation, we conclude that u^1 has the following form:

$$u^1(x,y) = \chi(y)\frac{du^0(x)}{dx} + \xi(x) \tag{2.18}$$

where $\chi(y)$ is the initial function of the terms inside the square brackets and $\xi(x)$ is the constant of integration due to y. From (2.18) and (2.16) it follows that

$$\sigma^0(x) = E(y)\left(1 + \frac{d\chi(y)}{dy}\right)\frac{du^0(x)}{dx}. \tag{2.19}$$

Differentiating (2.19) with respect to y, one concludes

$$\frac{d}{dy}\left[E(y)\left(1 + \frac{d\chi(y)}{dy}\right)\right] = 0, \qquad \text{on } Y \tag{2.20}$$

and $\chi(y)$ takes equal values on the opposite faces of Y (i.e. $\chi(0) = \chi(Y)$). Integrating (2.20) yields

$$E(y)\left(1 + \frac{d\chi(y)}{dy}\right) = a, \qquad (\ a \text{ is a constant }) \tag{2.21}$$

or

$$\frac{d\chi(y)}{dy} = \frac{a}{E(y)} - 1. \tag{2.22}$$

Integrating (2.22) it follows that

$$\chi(y) = \int_0^y \left(\frac{a}{E(\eta)} - 1 \right) d\eta + b \tag{2.23}$$

where η is the dummy variable of integration and b is a constant. Now using the boundary condition $\chi(0) = \chi(Y)$ yields

$$\int_0^Y \frac{a}{E(\eta)} d\eta - Y = 0 \tag{2.24}$$

or

$$a = 1/\left(\frac{1}{Y} \int_0^Y \frac{d\eta}{E(\eta)} \right). \tag{2.25}$$

Note that comparing (2.19) and (2.21) one can see that

$$\sigma^0(x) = a \frac{du^0(x)}{dx} \tag{2.26}$$

and substituting for a from (2.25) yields

$$\sigma^0(x) = 1/\left(\frac{1}{Y} \int_0^Y \frac{d\eta}{E(\eta)} \right) \frac{du^0(x)}{dx}. \tag{2.27}$$

By integrating (2.15) over the length of the period $(0, Y)$ and using fact (2) mentioned earlier, the following result is obtained

$$\frac{d\sigma^0(x)}{dx} + \overline{\gamma} = 0, \tag{2.28}$$

where $\overline{\gamma} = \frac{1}{Y} \int_Y \gamma(y) dy$ is the volumetric average of γ inside the base cell.

By studying (2.27) and (2.28), we realize that they are very similar to the equations of one dimensional elasticity for the homogeneous material and σ^0 and u^0 are independent of the microscopic scale y. The only difference is the elasticity coefficient which should be replaced by an homogenized one. Hence the problem can be summarized as :

$$\begin{cases} \sigma^0(x) = E^H du^0(x)/dx \\ d\sigma^0(x)/dx + \overline{\gamma} = 0 \end{cases} \tag{2.29}$$

where

$$E^H = 1/\left(\frac{1}{Y} \int_0^Y \frac{d\eta}{E(\eta)} \right) \tag{2.30}$$

is the *homogenized modulus of elasticity.*

To find the displacements, we follow the same process as for the homogeneous material bar problem. Combining the equations of (2.29) we obtain

$$\frac{\partial^2 u^0(x)}{\partial x^2} = -\frac{\overline{\gamma}}{E^H} .$$

By integrating twice and using the boundary conditions ($x = 0$; $u = 0$) and ($x = L$; $du/dx = 0$) we obtain

$$u(x) = -\frac{\overline{\gamma}}{E^H}\frac{x^2}{2} + \frac{\overline{\gamma}}{E^H}Lx.$$

Problem of heat conduction

The one dimensional heat conduction is very similar to the one dimensional elasticity problem. The governing equations, Fourier's law of heat conduction and the equation of heat balance, are

$$\begin{cases} q^\epsilon(x) = K^\epsilon \frac{dT^\epsilon(x)}{dx} \\ \frac{\partial q^\epsilon}{\partial x} + f = 0. \end{cases} \tag{2.31}$$

where q^ϵ is the heat flux, T^ϵ is the temperature and $K^\epsilon(x)$ is the conductivity coefficient. Following a very similar procedure to that used for the one dimensional elasticity problem the homogenized coefficient of heat conduction can be obtained as

$$K^H = 1/\left(\frac{1}{Y}\int_0^Y \frac{d\eta}{K(\eta)}\right)$$

which is as expected, the same as (2.30).

Similarly starting from the equations of heat conduction in the general three dimensional case and following the same procedure as for the one dimensional problem, the following results will be obtained [6]:

$$\begin{cases} \overline{q}_i(\mathbf{x}) = K^H_{ij}\frac{\partial \overline{T}(\mathbf{x})}{\partial x_j}, \\ \frac{\partial \overline{q}_i}{\partial x_i} + f = 0 \end{cases} \tag{2.32}$$

where

$$K^H_{ij} = \frac{1}{|\mathbf{Y}|}\left[\int_Y K(\mathbf{y})(\delta_{ij} + \frac{\partial \chi^j}{\partial y_i})d\mathbf{y}\right] \tag{2.33}$$

and $\chi^j(\mathbf{y})$ is the solution of the partial differential equation:

$$\frac{\partial}{\partial y_i}\left[K(\mathbf{y})(\delta_{ij} + \frac{\partial \chi^j}{\partial y_i})\right] = 0 \qquad \text{on } \mathbf{Y} \tag{2.34}$$

δ_{ij} is the Kronecker delta symbol and the boundary conditions are concluded from the periodicity, i.e. χ^j takes equal values on the opposite sides of the base cell. In (2.32), $\overline{q}$ and $\partial\overline{q}_i/\partial x_i$ are the volumetric average value of $q^0_i(x)$

and $\partial q_i^0/\partial x_i$ over $\mathbf{Y}$. The volumetric average of a quantity $a(\mathbf{x},\mathbf{y})$ over $\mathbf{Y}$ is defined by:

$$\overline{a}(\mathbf{x}) = \frac{1}{|\mathbf{Y}|}\int_Y a(\mathbf{x},\mathbf{y})d\mathbf{y}. \tag{2.35}$$

2.4 General boundary value problem

Many physical systems which do not change with time - sometimes called steady state problems - can be modelled by elliptic equations. As a general problem the divergent elliptic equation in a non-homogeneous medium with regular structure is now explained. Let $\Omega \subset I\!R^3$ be an unbounded medium tissued by parallelepiped unit cells $\mathbf{Y}$ whose material properties are determined by a symmetric matrix $a_{ij}(\mathbf{x},\mathbf{y}) = a_{ij}(\mathbf{y})$ where $\mathbf{y} = \mathbf{x}/\epsilon$ and $\mathbf{x} = (x_1, x_2, x_3)$ and the functions a_{ij} are periodic in the spatial variables $\mathbf{y} = (y_1, y_2, y_3)$. The boundary value problem to be dealt with is

$$\mathcal{A}^\epsilon u^\epsilon = f \qquad \text{in } \Omega \tag{2.36}$$

$$u^\epsilon = 0 \qquad \text{on } \partial\Omega \tag{2.37}$$

where the function f is defined in Ω and

$$\mathcal{A}^\epsilon = \frac{\partial}{\partial x_i}\left(a_{ij}(\mathbf{y})\frac{\partial}{\partial x_j}\right) \tag{2.38}$$

is the elliptical operator. The superscript ϵ is used to show the dependency of the operator and the solution to the characteristic inhomogeneity dimension.

Using a double-scale asymptotic expansion, the solution to (2.36) and (2.37) can be written as

$$u^\epsilon(\mathbf{x}) = u^0(\mathbf{x},\mathbf{y}) + \epsilon^1 u^1(\mathbf{x},\mathbf{y}) + \epsilon^2 u^2(\mathbf{x},\mathbf{y}) + \cdots \tag{2.39}$$

where functions $u^j(\mathbf{x},\mathbf{y})$ are $\mathbf{Y}$-periodic in $\mathbf{y}$. Recalling the rule of indirect differentiation (fact 3) yields

$$\mathcal{A}^\epsilon = \frac{1}{\epsilon^2}\mathcal{A}^1 + \frac{1}{\epsilon}\mathcal{A}^2 + \mathcal{A}^3 \tag{2.40}$$

where

$$\mathcal{A}^1 = \frac{\partial}{\partial y_i}\left(a_{ij}(\mathbf{y})\frac{\partial}{\partial y_j}\right) \quad ; \quad \mathcal{A}^3 = \frac{\partial}{\partial x_i}\left(a_{ij}(\mathbf{y})\frac{\partial}{\partial x_j}\right)$$

and

$$\mathcal{A}^2 = \frac{\partial}{\partial y_i}\left(a_{ij}(\mathbf{y})\frac{\partial}{\partial x_j}\right) + \frac{\partial}{\partial x_i}\left(a_{ij}(\mathbf{y})\frac{\partial}{\partial y_j}\right).$$

Applying (2.39) and (2.40) into (2.36) yields

$$\left(\epsilon^{-2}\mathcal{A}^1 + \epsilon^{-1}\mathcal{A}^2 + \mathcal{A}^3\right)\left(u^0 + \epsilon u^1 + \epsilon^2 u^2 + \cdots\right) = f, \tag{2.41}$$

and by equating terms with the same power of ϵ, we obtain

$$\mathcal{A}^1 u^0 = 0 \tag{2.42}$$

$$\mathcal{A}^1 u^1 + \mathcal{A}^2 u^0 = 0 \tag{2.43}$$

$$\mathcal{A}^1 u^2 + \mathcal{A}^2 u^1 + \mathcal{A}^3 u^0 = f \quad ; \cdots \tag{2.44}$$

If $\mathbf{x}$ and $\mathbf{y}$ are considered as independent variables these equations form a recurrent system of differential equations with the functions u^0, u^1 and u^2 parameterized by $\mathbf{x}$. Before proceeding to the analysis of this system, it is useful to notice to the following fact:

Fact (4). The equation

$$\mathcal{A}^1 u = F \qquad \text{in } \mathbf{Y} \tag{2.45}$$

for a $\mathbf{Y}$-periodic function u has a unique solution if

$$\overline{F} = \frac{1}{|Y|} \int_Y F d\mathbf{y} = 0 \tag{2.46}$$

where $|Y|$ denotes the volume of the base cell.

From this fact and using (2.42) it immediately follows that

$$u^0 = u(\mathbf{x}), \tag{2.47}$$

and by substituting into (2.43) we find

$$\mathcal{A}^1 u^1 = -\mathcal{A}^2 u^0 = -\frac{\partial a_{ij}(\mathbf{y})}{\partial y_i} \frac{\partial u(\mathbf{x})}{\partial x_j} \,. \tag{2.48}$$

As in the right hand side of (2.48) the variables are separated, the solution of this equation may be represented in the form

$$u^1(\mathbf{x}, \mathbf{y}) = \chi^j(\mathbf{y}) \frac{\partial u(\mathbf{x})}{\partial x_j} + \xi(\mathbf{x}) \tag{2.49}$$

where $\chi^j(\mathbf{y})$ is the $\mathbf{Y}$-periodic solution of the local equation

$$\mathcal{A}^1 \chi^j(\mathbf{y}) = \frac{\partial a_{ij}(\mathbf{y})}{\partial y_i} \qquad \text{in } \mathbf{Y}. \tag{2.50}$$

Now turning to (2.44) for u^2 and taking $\mathbf{x}$ as a parameter, it follows from fact (4) that (2.44) will have a unique solution if

$$-\frac{1}{|Y|} \int_Y (\mathcal{A}^2 u^1 + \mathcal{A}^3 u^0) d\mathbf{y} + f = 0, \tag{2.51}$$

which when combined with (2.49) results in the following homogenized (macroscopic) equation for $u(\mathbf{x})$

$$a_{ij}^H \frac{\partial^2 u(\mathbf{x})}{\partial x_i \partial x_j} = f \tag{2.52}$$

where the quantities

$$a_{ij}^H = \frac{1}{|Y|} \int_Y \left(a_{ij}(\mathbf{y}) + a_{ik}(\mathbf{y}) \frac{\partial \chi^j}{\partial y_k} \right) d\mathbf{y} \tag{2.53}$$

are the effective coefficients of the homogenized operator

$$\mathcal{A}^H = a_{ij}^H \frac{\partial^2}{\partial x_i \partial x_j} .$$

Thus it is demonstrated that the initial equation has been split into two different problems:

1. Determine $\chi^j(\mathbf{y})$ from (2.50) which is solved on the base cell.
2. Solve (2.52) on Ω with $u = 0$ on $\partial\Omega$. The homogenized coefficients a_{ij}^H are obtained from (2.53).

2.5 Elasticity problem in cellular bodies

So far the application of the homogenization theory in one dimensional elasticity and as a more general problem in elliptic partial differential equations has been discussed. For the sake of completeness, we now briefly explain the homogenization method for cellular media in weak form, which is suitable for the derivation of the finite element formulation, using the procedure and notation used by Guedes and Kikuchi [9]. This is the case applied in topological structural optimization by Bendsøe and Kikuchi [10-14].

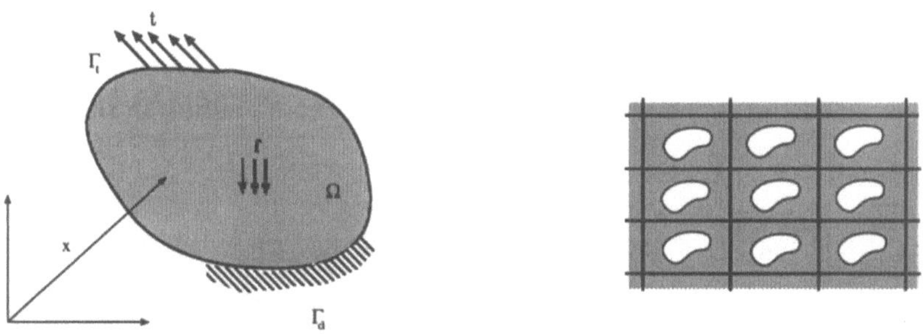

General elasticity problem   Cellular structure

Figure 2.6   Elasticity problem in a cellular body

Let us consider the elasticity problem constructed from a material with a porous body with a periodic cellular microstructure. Body forces **f** and tractions **t** are applied. See Figure 2.6. Ω is assumed to be an open subset of $\mathbb{R}^3$ with a smooth boundary on Γ comprising Γ_d (where displacements are prescribed) and Γ_t (the traction boundary). The base cell† of the cellular body $\mathbf{Y}$ is illustrated in Figure 2.7. $\mathbf{Y}$ is assumed to be an open rectangular parallelepiped in $I\!R^3$ defined by

$$\mathbf{Y} =]0, Y_1[\times]0, Y_2[\times]0, Y_3[,$$

with a hole v in it. The boundary of v is defined by s ($\partial v = s$) and is assumed to be sufficiently smooth and as a more general case the tractions $\mathbf{p}$ can also exist inside the holes. The solid part of the cell is denoted by $\mathbf{¥}$, therefore the solid part of the domain can be defined as

$$\Omega^\epsilon = \{\mathbf{x} \in \Omega \mid (\mathbf{y} = \mathbf{x}/\epsilon) \in \mathbf{¥}\}.$$

Also we define

$$S^\epsilon = \bigcup_{i=1}^{\text{all cells}} s_i.$$

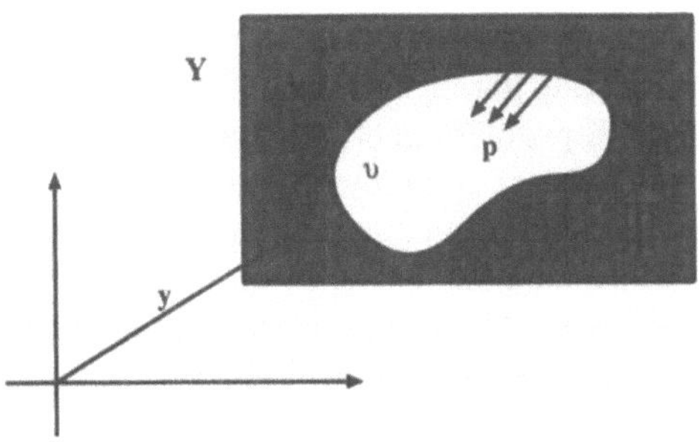

Figure 2.7 Base cell of the cellular body

It is assumed that none of the holes v_i intersect the boundary Γ, i.e.

$$\Gamma \cap S^\epsilon = \emptyset.$$

Now considering the stress-strain and strain-displacement relations

$$\sigma_{ij}^\epsilon = E_{ijkl}^\epsilon e_{kl}^\epsilon, \tag{2.54}$$

$$e_{kl}^\epsilon = \frac{1}{2}\left(\frac{\partial u_k^\epsilon}{\partial x_l} + \frac{\partial u_l^\epsilon}{\partial x_k}\right), \tag{2.55}$$

the virtual displacement equation can be constructed as:

† Having periodic microstructure does not mean that the form and composition of the base cell can not vary but the variations in the macroscopic scale are assumed to be smooth enough.

Find $\mathbf{u}^\epsilon \in \mathbf{V}^\epsilon$, such that

$$\int_{\Omega^\epsilon} E_{ijkl} \frac{\partial u_k^\epsilon}{\partial x_l} \frac{\partial v_i}{\partial x_j} d\Omega = \int_{\Omega^\epsilon} f_i^\epsilon v_i d\Omega + \int_{\Gamma_t} t_i v_i d\Gamma + \int_{S^\epsilon} p_i^\epsilon v_i dS \qquad \forall v \in \mathbf{V}^\epsilon. \tag{2.56}$$

where

$$\mathbf{V} = \left\{ v \in \left(H^1(\Omega^\epsilon) \right)^3 \text{ and } v|_{\Gamma_d} = \mathbf{0} \right\},$$

and H^1 is the Sobolev space†. The elastic constants of the solid are assumed to have symmetry and coercivity properties

$$E_{ijkl} = E_{jikl} = E_{ijlk} = E_{klij}$$

$$\exists \alpha > 0 : E_{ijkl}^\epsilon e_{ij} e_{kl} = \alpha e_{ij} e_{ij} \quad \forall e_{ij} = e_{ji}\,.$$

Now using the double-scale asymptotic expansion and fact (3), (2.56) becomes

$$\int_{\Omega^\epsilon} E_{ijkl} \left\{ \frac{1}{\epsilon^2} \frac{\partial u_k^0}{\partial y_l} \frac{\partial v_i}{\partial y_j} + \frac{1}{\epsilon} \left[\left(\frac{\partial u_k^0}{\partial x_l} + \frac{\partial u_k^1}{\partial y_l} \right) \frac{\partial v_i}{\partial y_j} + \frac{\partial u_k^0}{\partial y_l} \frac{\partial v_i}{\partial x_j} \right] + \right.$$

$$\left. \left[\left(\frac{\partial u_k^0}{\partial x_l} + \frac{\partial u_k^1}{\partial y_l} \right) \frac{\partial v_i}{\partial x_j} + \left(\frac{\partial u_k^1}{\partial x_l} + \frac{\partial u_k^2}{\partial y_l} \right) \frac{\partial v_i}{\partial y_j} \right] + \epsilon(\cdots) \right\} d\Omega$$

$$= \int_{\Omega^\epsilon} f_i^\epsilon v_i d\Omega + \int_{\Gamma_t} t_i v_i d\Gamma + \int_{S^\epsilon} p_i^\epsilon v_i ds \qquad \forall v \in \mathbf{V}_{\Omega \times \mathbf{Y}}. \tag{2.57}$$

where

$$\mathbf{V}_{\Omega \times \mathbf{Y}} = \{ v(\mathbf{x}, \mathbf{y}); (\mathbf{x}, \mathbf{y}) \in \Omega \times \mathbf{Y} \mid v(., \mathbf{y}) \ \mathbf{Y}\text{-- periodic };$$
$$v \ \text{ smooth enough and } v|_{\Gamma_d} = \mathbf{0} \}.$$

Similarly, we define $\mathbf{V}_\Omega$ and $\mathbf{V}_\mathbf{Y}$ as

$$\mathbf{V}_\Omega = \{ v(\mathbf{x}) \text{ defined in } \Omega \mid v \text{ smooth enough and } v|_{\Gamma_d} = \mathbf{0} \}.$$

$$\mathbf{V}_\mathbf{Y} = \{ v(\mathbf{y}) \text{ defined in } \mathbf{Y} \mid v(\mathbf{y}), \ \mathbf{Y}\text{-- periodic and smooth enough} \}.$$

Introducing the following facts:

---

† $H^1(\Omega^\epsilon)$ is defined as

$$H^1(\Omega^\epsilon) = \left\{ w(\mathbf{x}) \mid w(\mathbf{x}) \in L_2(\Omega^\epsilon) \text{ and } \frac{\partial w(\mathbf{x})}{\partial x_i} \in L_2(\Omega^\epsilon) \right\}$$

where

$$L_2(\Omega^\epsilon) = \left\{ w(\mathbf{x}) \mid \int_{\Omega^\epsilon} (w(\mathbf{x}))^2 \, d\mathbf{x} < \infty \quad \text{and } \mathbf{x} \in \Omega^\epsilon \right\},$$

which assures the integrability of the functions and their derivatives.

Fact (5). For a Y-periodic function $\Psi(\mathbf{y})$ when $\epsilon \to 0$ we have

$$\int_{\Omega^\epsilon} \Psi\left(\frac{\mathbf{x}}{\epsilon}\right) d\Omega = \frac{1}{|Y|}\int_\Omega \int_{¥} \Psi(\mathbf{y}) dY d\Omega \,, \tag{2.58}$$

$$\int_{S^\epsilon} \Psi\left(\frac{\mathbf{x}}{\epsilon}\right) d\Omega = \frac{1}{\epsilon|Y|}\int_\Omega \int_{s} \Psi(\mathbf{y}) ds d\Omega \,, \tag{2.59}$$

and assuming that the functions are all smooth so that when $\epsilon \to 0$ all integrals exist and by equating the terms with the same power of ϵ we obtain

$$\frac{1}{|Y|}\int_\Omega \int_{¥} E_{ijkl}\frac{\partial u_k^0}{\partial y_l}\frac{\partial v_i}{\partial y_j} dY d\Gamma = 0 \qquad \forall v \in \mathbf{V}_{\Omega \times ¥}, \tag{2.60}$$

$$\int_\Omega \left\{ \frac{1}{|Y|}\int_{¥} E_{ijkl}\left[\left(\frac{\partial u_k^0}{\partial x_l} + \frac{\partial u_k^1}{\partial y_l}\right)\frac{\partial v_i}{\partial y_j} + \frac{\partial u_k^0}{\partial y_l}\frac{\partial v_i}{\partial x_j}\right] dY \right\} d\Omega$$

$$= \int_\Omega \left(\frac{1}{|Y|}\int_s p_i v_i dS\right) d\Omega \qquad \forall v \in \mathbf{V}_{\Omega \times ¥}, \tag{2.61}$$

$$\int_\Omega \left\{ \frac{1}{|Y|}\int_{¥} E_{ijkl}\left[\left(\frac{\partial u_k^0}{\partial x_l} + \frac{\partial u_k^1}{\partial y_l}\right)\frac{\partial v_i}{\partial x_j} + \left(\frac{\partial u_k^1}{\partial x_l} + \frac{\partial u_k^2}{\partial y_l}\right)\frac{\partial v_i}{\partial y_j}\right] dY \right\} d\Omega$$

$$= \int_\Omega \left(\frac{1}{|Y|}\int_{¥} f_i v_i dY\right) d\Omega + \int_{\Gamma_t} t_i v_i d\Gamma \qquad \forall v \in \mathbf{V}_{\Omega \times ¥}. \tag{2.62}$$

Now, as v is an arbitrary function we choose $v = v(\mathbf{y})$ (i.e. $v \in \mathbf{V}_{¥}$). Then, integrating by parts, applying the divergence theorem to the integral in ¥, and using periodicity from (2.60) we obtain

$$\frac{1}{|Y|}\int_\Omega \left\{ \int_{¥}\left[-\frac{\partial}{\partial y_j}\left(E_{ijkl}\frac{\partial u_k^0}{\partial y_l}\right)\right] v_i dY + \int_s E_{ijkl}\frac{\partial u_k^0}{\partial y_l} n_j v_i dS \right\} d\Omega = 0 \qquad \forall v. \tag{2.63}$$

v being arbitrary results in

$$-\frac{\partial}{\partial y_j}\left(E_{ijkl}\frac{\partial u_k^0}{\partial y_l}\right) = 0 \qquad \forall \mathbf{y} \in ¥, \tag{2.64}$$

$$E_{ijkl}\frac{\partial u_k^0}{\partial y_l} n_j = 0 \qquad \text{on } s. \tag{2.65}$$

Considering fact (4) and (2.64) it is concluded that

$$\mathbf{u^0}(\mathbf{x}, \mathbf{y}) = \mathbf{u^0}(\mathbf{x}) \,. \tag{2.66}$$

This means that the first term of the asymptotic expansion only depends on the macroscopic scale $\mathbf{x}$.

Now, as v is an arbitrary function if we choose $v = v(\mathbf{x})$ (i.e. v is only a function of $\mathbf{x}$) then from (2.61) it is concluded that

$$\int_{\Omega}\left(\frac{1}{|Y|}\int_{s} p_i dS\right) v_i(\mathbf{x}) d\Omega = 0 \qquad \forall v \in \mathbf{V}_{\Omega}, \tag{2.67}$$

which implies that

$$\int_{s} p_i(\mathbf{x},\mathbf{y}) dS = 0. \tag{2.68}$$

This means that the applied tractions have to be self-equilibrating. So the possible applied tractions are restricted.

On the other hand, introducing (2.66) into (2.61) and choosing $v = v(\mathbf{y})$ yields

$$\int_{\yen} E_{ijkl}\left(\frac{\partial u_k^0}{\partial x_l} + \frac{\partial u_k^1}{\partial y_l}\right)\frac{\partial v_i(\mathbf{y})}{\partial y_j} dY = \int_{s} p_i v_i dS \qquad \forall v \in \mathbf{V}_{\yen}. \tag{2.69}$$

Integrating by parts, using the divergence theorem and applying the periodicity conditions on the opposite faces of Y, it follows from (2.69) that

$$-\int_{\yen}\frac{\partial}{\partial y_j}\left[E_{ijkl}\left(\frac{\partial u_k^0(\mathbf{x})}{\partial x_l} + \frac{\partial u_k^1}{\partial y_l}\right)\right] v_i dY + \int_{s} E_{ijkl}\left(\frac{\partial u_k^0(\mathbf{x})}{\partial x_l} + \frac{\partial u_k^1}{\partial y_l}\right) v_i n_j dS$$

$$= \int_{s} p_i v_i dS \qquad \forall v \in \mathbf{V}_{\yen}. \tag{2.70}$$

Since v is arbitrary, it is concluded that

$$-\frac{\partial}{\partial y_j}\left(E_{ijkl}\frac{\partial u_k^1}{\partial y_l}\right) = \frac{\partial}{\partial y_j}\left(E_{ijkl}\frac{\partial u_k^0(\mathbf{x})}{\partial x_l}\right) \qquad \text{on } \yen, \tag{2.71}$$

$$E_{ijkl}\frac{\partial u_k^1}{\partial y_l} = -E_{ijkl}\frac{\partial u_k^0(\mathbf{x})}{\partial x_l} n_j + p_i \qquad \text{on } s. \tag{2.72}$$

Now considering (2.62) and choosing $v = v(\mathbf{x})$ results in a statement of equilibrium in the macroscopic level:

$$\int_{\Omega}\left[\frac{1}{|Y|}\int_{\yen} E_{ijkl}\left(\frac{\partial u_k^0}{\partial x_l} + \frac{\partial u_k^1}{\partial y_l}\right) dY\right]\frac{\partial v_i(\mathbf{x})}{\partial x_j} d\Omega$$

$$= \int_{\Omega}\left(\frac{1}{|Y|}\int_{\yen} f_i dY\right) v_i(\mathbf{x}) d\Omega + \int_{\Gamma_t} t_i v_i(\mathbf{x}) d\Gamma \qquad \forall v \in \mathbf{V}_{\Omega}. \tag{2.73}$$

If in (2.62) we assume that $v = v(\mathbf{y})$ leads to

$$\int_{\Omega}\left[\frac{1}{|Y|}\int_{\yen} E_{ijkl}\left(\frac{\partial u_k^1}{\partial x_l} + \frac{\partial u_k^2}{\partial y_l}\right)\frac{\partial v_i(\mathbf{y})}{\partial y_j} dY\right] d\Omega = \int_{\Omega}\left(\frac{1}{|Y|}\int_{\yen} f_i v_i(\mathbf{y}) dY\right) d\Omega$$

$$\forall v \in \mathbf{V}_{\yen}. \tag{2.74}$$

or equivalently

$$\int_{\text{¥}} E_{ijkl}\left(\frac{\partial u_k^1}{\partial x_l}+\frac{\partial u_k^2}{\partial y_l}\right)\frac{\partial v_i(\mathbf{y})}{\partial y_j}dY=\int_{\text{¥}} f_i v_i(\mathbf{y})dY \qquad \forall v\in \mathbf{V}_{\text{¥}}, \tag{2.75}$$

which represents equilibrium of the base cell in the microscopic level.

The procedure followed so far can be applied for higher terms of the expansion. However, in this case the first order terms are enough. The macroscopic mechanical behaviour is represented by $\mathbf{u}^0$ and $\mathbf{u}^1$ represents the microscopic behaviour.

As we have noticed earlier, our goal is to find the homogenized elastic constants such that the equilibrium equation (or equivalently the equation of virtual displacements) can be constructed in the macroscopic system of coordinates. These homogenized constants should be such that the corresponding equilibrium equation reflects the mechanical behaviour of the microstructure of the cellular material without explicitly using the parameter ϵ. To accomplish this we consider (2.69) once again. As this equation is linear with respect to $\mathbf{u}^0$ and $\mathbf{p}$, we consider the two following problems:

(i) Let $\chi^{kl} \in V_{\text{¥}}$ be the solution of

$$\int_{\text{¥}} E_{ijpq}\frac{\partial \chi_p^{kl}}{\partial y_q}\frac{\partial v_i(\mathbf{y})}{\partial y_j}dY=\int_{\text{¥}} E_{ijkl}\frac{\partial v_i(\mathbf{y})}{\partial y_j}dY \qquad \forall v\in \mathbf{V}_{\text{¥}}, \tag{2.76}$$

(ii) and let $\Psi \in \mathbf{V}_{\text{¥}}$ be the solution of

$$\int_{\text{¥}} E_{ijkl}\frac{\partial \Psi_k}{\partial y_l}\frac{\partial v_i(\mathbf{y})}{\partial y_j}dY=\int_{S} p_i v_i(\mathbf{y})dY \qquad \forall v\in \mathbf{V}_{\text{¥}} \tag{2.77}$$

where $\mathbf{x}$ plays the role of a parameter.

It can be shown that the solution $\mathbf{u}^1$ will be

$$u_i^1=-\chi_i^{kl}(\mathbf{x},\mathbf{y})\frac{\partial u_k^0(\mathbf{x})}{\partial x_l}-\Psi_i(\mathbf{x},\mathbf{y})+\tilde{u}_i^1(\mathbf{x}), \tag{2.78}$$

where $\tilde{u}_i^1$ are arbitrary constants of integration in $\mathbf{y}$.

Introducing (2.78) into (2.73) yields

$$\begin{aligned}\int_\Omega\left[\frac{1}{|Y|}\int_{\text{¥}}\left(E_{ijkl}-E_{ijpq}\frac{\partial\chi_p^{kl}}{\partial y_q}\right)dY\right]\frac{\partial u_k^0(\mathbf{x})}{\partial x_l}\frac{\partial v_i(\mathbf{x})}{\partial x_j}d\Omega\\ =\int_\Omega\left(\frac{1}{|Y|}\int_{\text{¥}}E_{ijkl}\frac{\partial\Psi_k}{\partial y_l}dY\right)\frac{\partial v_i(\mathbf{x})}{\partial x_j}d\Omega+\int_\Omega\left(\frac{1}{|Y|}\int_{\text{¥}}f_i dY\right)v_i(\mathbf{x})d\Omega\\ +\int_{\Gamma_t}t_i v_i(\mathbf{x})d\Gamma \qquad \forall v\in \mathbf{V}_\Omega.\end{aligned} \tag{2.79}$$

Now denoting

$$E_{ijkl}^H(\mathbf{x})=\frac{1}{|Y|}\int_{\text{¥}}\left(E_{ijkl}-E_{ijpq}\frac{\partial\chi_p^{kl}}{\partial y_q}\right)dY, \tag{2.80}$$

$$\tau_{ij}(\mathbf{x}) = \int_{\mathbf{Y}} E_{ijkl} \frac{\partial \Psi_k}{\partial y_l} dY, \tag{2.81}$$

and

$$b_i(\mathbf{x}) = \frac{1}{|Y|} \int_{\mathbf{Y}} f_i dY, \tag{2.82}$$

(2.79) can be written as

$$\int_{\Omega} E^H_{ijkl} \frac{\partial u^0_k(\mathbf{x})}{\partial x_l} \frac{\partial v_i(\mathbf{x})}{\partial x_j} d\Omega = \int_{\Omega} \tau_{ij}(\mathbf{x}) \frac{\partial v_i(\mathbf{x})}{\partial x_j} d\Omega$$
$$+ \int_{\Omega} b_i(\mathbf{x}) v_i(\mathbf{x}) d\Omega + \int_{\Gamma_t} t_i(\mathbf{x}) v_i(\mathbf{x}) d\Gamma \qquad \forall v \in \mathbf{V}_{\Omega}. \tag{2.83}$$

This is very similar to the equation of virtual displacement (2.56) and it represents the macroscopic equilibrium. E^H_{ijkl} defined by (2.80) represents the homogenized elastic constants. τ_{ij} are average 'residual' stresses within the cell due to the tractions $\mathbf{p}$ inside the holes and b_i are the average body forces.

As we notice the microscopic and macroscopic problems are not coupled and the solution of the elasticity problem can be summarized as:

1. Find χ and Ψ within the base cell by solving the integral equations (2.76) and (2.77) on the base cell.
2. Find E^H_{ijkl}, τ_{ij} and b_i by using (2.80), (2.81) and (2.82).
3. Construct (2.83) in macroscopic coordinates.

If the whole domain of the cellular material comprises a uniform cell structure, as well as uniform applied tractions on the boundaries of the holes of the cells, then it is only necessary to solve the microscopic equations (2.76) and (2.77) once. Otherwise these equations must be solved for every point $\mathbf{x}$ of Ω.

References

[1] Kalamkarov A. L., *Composite and Reinforced Elements of Construction.* John Wiley & Sons, Chichester (1992)

[2] Sanchez-Palencia E., Non-homogenous media and vibration theory, *Lecture Notes in Physics*, **127** (1980)

[3] Benssousan A., Lions J.L. and G. Papanicoulau, *Asymptotic analysis for periodic structures.* North Holland, Amsterdam (1978)

[4] Cioranescu D. and Paulin J.S.J., Homogenization in open sets with holes, *Journal of Math. Analysis and Appl.*, (71), 590–607 (1979)

[5] Oleinik O.A., On homogenization problems, in *Trends and application of pure mathematics in mechanics*, Springer, Berlin (1984)

[6] Caillerie D., Homogenization of periodic media tissued composite materials, Tech. rep., Institute of Mechanics, Grenoble, France

[7] Bourgat J.F., Numerical experiments of the homogenization method for operators with periodic coefficients, *Lecture Notes in Mathematics*, **704**, 330–356 (1979)

[8] Lene F. and Duvaut G., Resultats d'isotropie pour des milieux homogeneises, *C.R. Acad. Sc. Paris*, **7 293, Serie II**, 477–480 (1981)

[9] Guedes J.M. and Kikuchi N., Pre and post processing for materials based on the homogenization method with adaptive finite element methods, *Comp. Meth. Appl. Mech. Eng.*, **83**, 143–198 (1990)

[10] Bendsøe M.P. and Kikuchi N., Generating optimal topologies in structural design using homogenization method, *Comp. Meth. Appl. Mech. Eng.*, **71**, 197–224 (1988)

[11] Bendsøe M.P., Optimal shape design as a material distribution problem, *Structural Optimization*, **1**, 193–202 (1989)

[12] Bendsøe M.P., Díaz A.R. and Kikuchi N., Topology and generalized layout optimization of elastic structures, in *Topology design of structures*, edited by Bendsøe M.P. and Mota Soares C. A., pp. 159–205. Kluwer Academic Publishers (1993)

[13] Suzuki K. and Kikuchi N., A homogenization method for shape and topology optimization, *Comp. Meth. Appl. Mech. Eng.*, **93**, 291–318 (1991)

[14] Jog C.S., Haber R.B. and Bendsøe M.P., Topology design with optimized, self-adaptive materials, Tech. Rep. DCAMM 457, Technical University of Denmark (1993)

3

SOLUTION OF HOMOGENIZATION EQUATIONS FOR TOPOLOGY OPTIMIZATION

In this chapter motives for using the homogenization theory for topological structural optimization are briefly explained. Different material models are described and the analytical solution of the homogenization equations, derived in the last section of Chapter 2, for the so called 'rank laminate composites' is presented. The finite element formulation is explained for the material model based on a microstructure consisting of an isotropic material with rectangular voids. Using the periodicity assumption, the boundary conditions are derived and the homogenization equation is solved. The results to be used in topology optimization are presented.

3.1 Introduction

Structural topology optimization has the complex features of both size and shape optimization problems. One of the drawbacks of conventional shape optimization by the boundary variation method is that final designs are topologically equivalent to the initial ones. Normally these techniques require several finite element re-meshings in the optimization process. Furthermore, the difficulty of changing the topology during the scheme increases the complexity of the problem. In topological structural optimization we need to be able to create new holes and this can not be achieved in the boundary variation method. In addition, unlike the conventional shape optimization methods where the design boundary can be described by a set of simple geometrical segments (i.e. lines, circular arcs, elliptic arcs and splines), it is not easy to define the structural topology optimization problem by using a finite number of parameters.

Studying the nature of the topological structural shape optimization problem suggests a specific and different sort of parameterization from the outset: that is the natural formation of an infinite number of holes of varying size which implies the existence of a special microstructure. It has already been proven that in some cases, having a finite number of design variables does not lead to the optimum solution. For example, in the problem of thickness optimization of a linearly elastic plate, with mean compliance considered as the objective function, Cheng and Olhoff [1,2] demonstrated that it is not realistic to define this problem by using a fixed, finite number of parameters and basic functions, as the true optimum problem consists of many discrete ribs of various sizes. This problem has also been studied by Lurie *et al* [3] using the theory of G-closure. They concluded that it is necessary to introduce the relaxed form of the state equation by using an orthotropic plate that may possess a microstructure. This was also confirmed by Rozvany, Olhoff, Bendsøe *et al* [4] and Ong, Rozvany and Szeto [5]. Kohn and Strang [6] established that in the context of plastic design for torsion of a cross section within a square area, the generalized shape optimization (i.e. both shape and topology) may yield three types of regions: solid, empty and porous, where the porous regions comprise some material with infinitesimal cavities. See Reference [7]. Also by introducing a microstructure to the material model [8], they could obtain a well-posed, relaxed formulation for a two dimensional heat conduction problem. The main advantage of this relaxed formulation is that the solution with the finite element method is no longer mesh-dependent.

Structural shape optimization in general, can be thought of as the determination of the optimal spatial material distribution. In other words, for a given set of loads and boundary conditions, the problem is how to redistribute the material in order to minimize the objective function, for example, the mean compliance. Therefore the general shape optimization problem can be considered as a point-wise material/no material one which does not need to be represented by the shape parameters or basic shape functions. However, implementation of this on-off approach to an optimization problem requires the use of discrete optimization algorithms. Such an approach would be unstable, unless composite materials were introduced. 'In general, existence of solutions can not be expected unless the problem is turned into a material distribution one, using composite materials' (Bendsøe [9]).

Introducing a material density function by considering a composite consisting of an infinite number of infinitely small holes which are periodically distributed, the complex nature of the structural topological optimization problem can be converted to a sizing problem. In fact using the idea of using a cellular body with a periodic microstructure, moves the on-off nature of the problem from the macroscopic scale to the microscopic scale [9]. In mathematical terms the introduction of microstructures corresponds to a relaxation of the variational problem that can be established for the optimization problem [8, 10].

There are many ways to introduce the above mentioned microstructures and these can be divided into two categories: rank laminate composite methods and microcells with internal voids that will be defined in the next section. The theory of homogenization is used to determine the macroscopic mechanical properties of these materials. In the case of rank materials the homogenization equation can be solved analytically but for the other voided microcells it usually has to be solved by numerical methods such as finite elements.

In practice, after choosing the reference domain and finite element discretization, it is assumed that each element consists of a cellular material with a specific microstructure, and the geometrical parameters of these microstructures are the design variables of the optimization problem. It is noted that the problem is solved in a fixed domain so that the finite element model used in the analysis does not need to be altered during the optimization algorithm.

An alternative direct approach to the discrete optimization problem of material distribution exists which substitutes the on-off character of the problem by a suitable differentiable approximation called an 'artificial density function'. These are usually referred to as artificial materials. In this book microcells with rectangular voids, the so called rank-2 materials and the artificial materials are considered.

3.2 Material models

As mentioned in the introduction, the main idea of solving a class of shape optimization problems involving topology is to introduce an 'infinite' number of microscale voids to form a porous medium. The optimization problem is defined in such a way that the geometry parameters of these voids are the design variables. If a portion of the medium consists only of voids, material is not placed over that area. On the other hand, if there is no porosity at some portion, a solid structure needs to be placed at that location. Some common ways of creating these microstructures are now introduced.

3.2.1 Rectangular microscale voids

In choosing the microstructure, one of the important features that should be considered is that it should allow the density of material to cover the whole range of values from zero (void) to one (solid). For example, circular holes (spherical in 3D) do not allow this, because they can not provide a complete void. In addition they should fit the periodicity assumption and their geometry should be defined by a very few parameters (as these are the design variables of the optimization algorithm). Square cells with centrally-placed rectangular holes are one of the simplest means of achieving this objective (Figure 3.1). In three dimensional space this microcell will be a cubic cell with a rectangular parallelepiped hole in it.

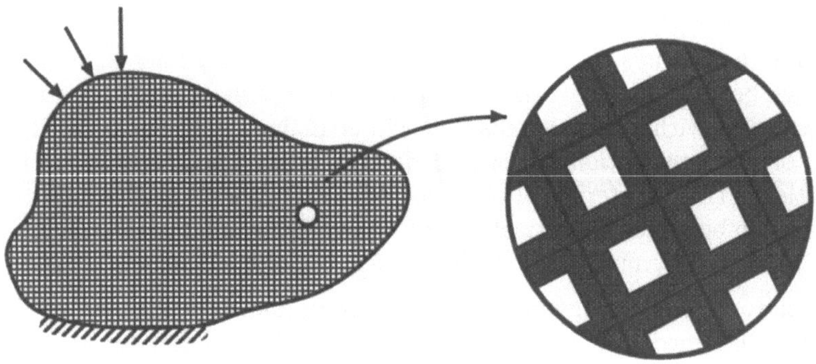

Figure 3.1 Microcells with rectangular holes

Another possibility, for example, is the generalized ellipsoid [11] defined by

$$\left(\frac{y_1}{a}\right)^m + \left(\frac{y_2}{b}\right)^m + \left(\frac{y_3}{c}\right)^m = 1,$$

where a, b and c are the principal radii of the ellipsoid and m is a large number.

If the dimension of the square cell is denoted by ϵ which is a very small positive number, then the size of the hole will be ϵa and ϵb. Figure 3.2 illustrates this unit cell in the microscopic coordinates. Using this model, the area occupied by the solid material is given by

$$\Omega_s = \int_\Omega (1 - ab) d\Omega, \tag{3.1}$$

where $0 \leq a \leq 1$, $0 \leq b \leq 1$, Ω is the design domain and Ω_s denotes the solid part of it.

In general, the microscale perforations in the cellular body, with respect to the coordinate axes, can have different orientations. Since this orientation θ will affect the properties of the elastic constitutive matrix, it has to be taken into account in the formulation. As each point $\mathbf{x} \in \Omega$ has its own a, b and θ values, these parameters can be considered as distributed functions of the position vector $\mathbf{x}$ and these are the design variables of the relaxed optimization problem. According to the theory, the functions $a = a(\mathbf{x})$, $b = b(\mathbf{x})$ and $\theta = \theta(\mathbf{x})$ are assumed to be smooth enough, for example, $a, b, \theta \in H^1(\Omega)$. On the other hand, to be able to construct the optimization formulation, these functions have to be represented by discrete approximations, otherwise there will be an infinite number of design variables. In practice, these functions are approximated by constant functions in each finite element of the discretized model of the design domain. Hence, a constant microstructure is assumed inside each element and the homogenized elasticity matrix will therefore also be constant in each finite element. Thus, if the design domain Ω is decomposed into N finite elements, then we will have $3N$ design variables, (i.e. (a_i, b_i, θ_i), $\quad i = 1, 2, \cdots, N$).

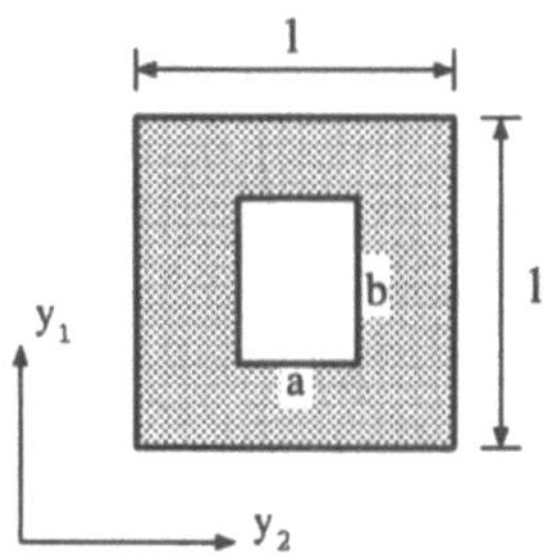

Figure 3.2 *Unit cell with rectangular hole in microscopic coordinates*

Assuming that the solid part of the microstructure is an isotropic material, because of the rectangular perforations the cellular body becomes an orthotropic one. For the two dimensional (plane stress/strain) elastic problem the constitutive law will be (see Appendix A)

$$\begin{Bmatrix} \sigma_{11} \\ \sigma_{22} \\ \sigma_{12} \end{Bmatrix} = \begin{bmatrix} D_{11} & D_{12} & 0 \\ D_{12} & D_{22} & 0 \\ 0 & 0 & D_{66} \end{bmatrix} \begin{Bmatrix} \epsilon_{11} \\ \epsilon_{22} \\ 2\epsilon_{12} \end{Bmatrix} \tag{3.2}$$

where σ_{ij} $\quad (i, j = 1, 2)$, are the stresses and ϵ_{ij} denotes the strains. D_{ij} are the elements of the reduced stiffness matrix of material.

It is noted that for the cellular material the elements of the matrix of elastic moduli $\mathbf{D}$ are functions of a, b and θ, so that

$$\mathbf{D} = \mathbf{D}(a, b, \theta). \tag{3.3}$$

The dependency on a and b is calculated by asymptotic homogenization theory and one can find the dependency on θ by using the well known frame rotation formula

$$\mathbf{D} = \mathbf{R}^T(\theta)\mathbf{D}^H(a, b)\mathbf{R}(\theta), \tag{3.4}$$

where **R** is the matrix of rotation. This issue will be discussed later. We also note that the density function ρ is a function of a and b

$$\rho = \rho(a, b) = (1 - ab)\rho_s, \tag{3.5}$$

where ρ_s is the density of the solid. Similarly, for inertia body forces we can write

$$\mathbf{f} = \mathbf{f}(a, b) = \rho_s(1 - ab)\gamma, \tag{3.6}$$

where γ is a known vector.

3.2.2 Ranked layered material cells

Layered materials are another type of microstructure that can be applied in order to produce a relaxed form of the topological structural optimization problem. Each cell of this periodic microstructure is constructed from layers of different materials and voids.

The so called rank-1 material is constructed of alternating layers of solid materials and voids. In practice, to avoid singularity in the stiffness matrix of the structure, instead of voids a very soft (flexible) material is used. The relative densities of the solid and soft layers are respectively denoted by γ and $1 - \gamma$. See Figure 3.3.

This material can be used to construct the higher ranks of layered composites. For example, the rank-2 composite is constructed by repeating layers of the solid and the rank-1 composite with relative densities of μ and $1 - \mu$, respectively. The direction of the layers of different ranks are orthogonal to each other. The characteristic width of these layers are of different orders. In other words, the real width of a unit layer (cell) of rank-1 material in comparison to the real width of rank-2 material is a very small number.

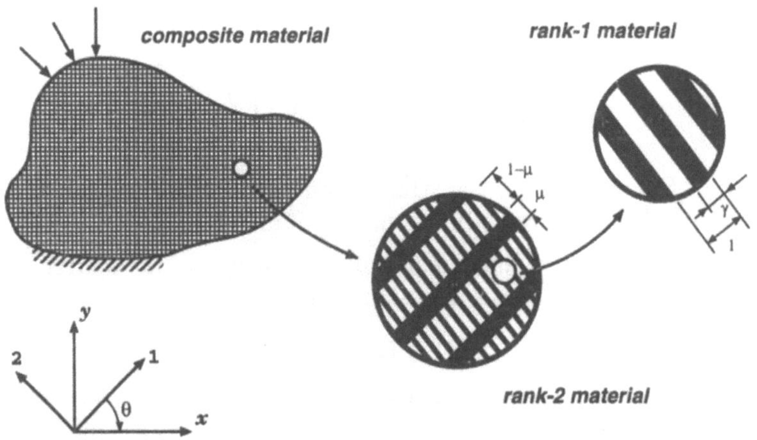

Figure 3.3 Construction of the rank-2 layered material

Studies on the effective material properties of composite mixtures of two different materials have shown that in plane elasticity problems, the rank-2 composites yield the stiffest material [12-15].

As for the case of microstructures with rectangular voids, the elements of the matrix of elasticity coefficients for rank-2 materials, are functions of three parameters: γ, μ and the orientation angle θ; see Figure 3.3, so that

$$\mathbf{D} = \mathbf{D}(\gamma, \mu, \theta). \tag{3.7}$$

The volume occupied by the solid (neglecting the very soft material) is

$$\Omega_s = \int_\Omega (\mu + \gamma - \mu\gamma) d\Omega \tag{3.8}$$

and the density of the composite can be written as

$$\rho = \rho(\gamma, \mu) = (\gamma + \mu - \gamma\mu)\rho_s \tag{3.9}$$

where ρ_s is the density of solid and $0 \leq \gamma, \mu \leq 1$. It is noted that by changing the values of γ and μ it is possible to cover the complete range of cell relative densities from zero (void) to one (solid).

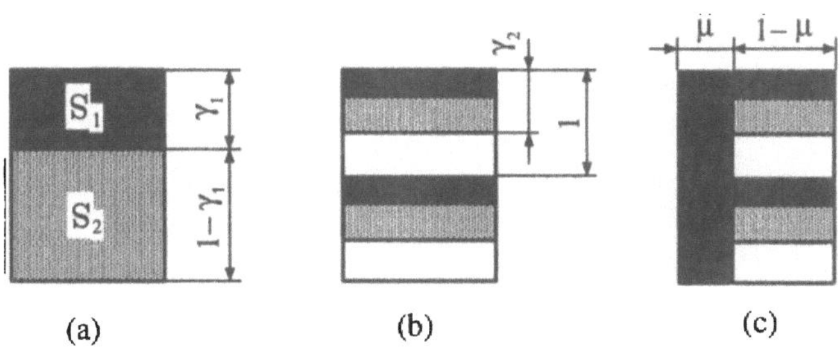

Figure 3.4 *Construction of a rank-2 bi-material composite*

Olhoff *et al* [16] and Thomsen [17] have used a bi-material rank-2 composite for the topology optimization problem. This model makes it possible to generate typical sandwich structures. The rank-1 bi-material is constructed from two isotropic materials: one with a relatively large stiffness, and one with a softer material, and a very soft material which represents the voids. The rank-2 material is constructed by using the stiffest material in one direction and the rank-1 composite layers in the perpendicular direction. Figure 3.4 illustrates the different levels of construction. At the first level shown in Figure 3.4(a), the composite is modeled by using isotropic stiff materials S_1 and S_2 with concentrations given by γ_1 and $1 - \gamma_1$ respectively. In the second level, the above mentioned material is compounded with a very soft material (representing the void) with relative concentrations of γ_2 and $1 - \gamma_2$, as shown in Figure 3.4(b). At the third level, the rank-2 composite is constructed by using material S_1

(the stiffest one) and the composite in Figure 3.4(b) with relative concentrations μ and $1-\mu$, respectively (Figure 3.4(c)). In this case the density at each point of the composite is a function of γ_1, γ_2 and μ and the densities of the stiff materials ρ_{s_1} and ρ_{s_2}, and has the form

$$\rho = \mu\rho_{s_1} + (1-\mu)\gamma_1\gamma_2\rho_{s_1} + (1-\mu)(1-\gamma_1)\gamma_2\rho_{s_2}. \tag{3.10}$$

For the volume of the materials we can write

$$\Omega_{s_1} = \int_\Omega [\mu + (1-\mu)\gamma_1\gamma_2]\, d\Omega \quad \text{and} \quad \Omega_{s_2} = \int_\Omega [(1-\mu)(1-\gamma_1)\gamma_2]\, d\Omega. \tag{3.11}$$

It is noted that the elements of the constitutive matrix in this case are functions of γ_1, γ_2, μ and θ, so that

$$\mathbf{D} = \mathbf{D}(\gamma_1, \gamma_2, \mu, \theta), \tag{3.12}$$

where θ is the angle of orientation of layers with respect to the main coordinates. By using this model for structural topology optimization in every element we have four design variables.

3.2.3 Artificial materials

By considering the general shape and topology optimization of structures as a material distribution problem, the structure can be described by a discrete function χ, defined at each point $\mathbf{x}$ as

$$\chi(\mathbf{x}) = \begin{cases} 1 & \text{if} \quad \mathbf{x} \in \Omega_s & \text{material} \\ 0 & \text{if} \quad \mathbf{x} \in \Omega \backslash \Omega_s & \text{no material.} \end{cases} \tag{3.13}$$

Assuming isotropy for the solid part of the structure, we can write

$$\rho(\mathbf{x}) = \chi(\mathbf{x})\rho^0, \tag{3.14}$$

and

$$\mathbf{D}(\mathbf{x}) = \chi(\mathbf{x})\mathbf{D}^0, \tag{3.15}$$

where ρ^0 and $\mathbf{D}^0$ are respectively the density and elasticity matrix of the homogeneous solid.

For the numerical solution of the optimization problem, the indicator function $\chi(\mathbf{x})$ has to be discretized and this yields an integer programming problem where $\chi(\mathbf{x}^e)$ within each element e are the design variables. But this formulation, apart from being expensive, is not well posed [8]. This is partly because of strong dependency of the optimization results on the chosen discretization and, furthermore, the integer formulation comprises many artificial local minima [18].

The very easy way to relax the problem is to replace the discrete value parameter $\chi(\mathbf{x})$ with a continuous one $\xi(\mathbf{x})$, so that

$$\rho(\mathbf{x}) = \xi(\mathbf{x})\rho^0 \tag{3.16}$$

and

$$\mathbf{D}(\mathbf{x}) = \xi(\mathbf{x})\mathbf{D}^0, \tag{3.17}$$

where $0 \leq \xi(\mathbf{x}) \leq 1$ and $\mathbf{x} \in \Omega$. Note that according to (3.16) the volume of material, V, is obtained by

$$V = \int_\Omega \xi(\mathbf{x}) d\Omega. \tag{3.18}$$

Although (3.16) and (3.17) yield a relaxed optimization algorithm, it results in some porous areas in the optimum structure. From an engineering point of view, it is more practical to obtain a solution which only consists of solid and void regions. Hence it may be desirable to suppress the porous areas by penalizing the intermediate values for $\xi(\mathbf{x})$. This idea is explained by Rozvany *et al* by introducing the fabrication cost of voids [7, 19, 20]. Also it can easily be provided by changing (3.17) to the form

$$\mathbf{D}(\mathbf{x}) = \xi(\mathbf{x})^\mu \mathbf{D}^0, \tag{3.19}$$

where $\mu > 1$ and is usually between 3 and 9 [9, 21, 22]. As a general rule, larger μ values result in fewer perforated regions. Note that since the volume of the structure is linear in ξ, the intermediate values give a little stiffness at an undesirable cost, and are therefore less likely to appear in the optimal layout. This problem is discussed with more details in Section 5.2.2.

The artificial density function ξ can also be related to some geometrical parameters to create some sort of artificial microstructures. For example, to make analogy to the idea of a cellular body consisting of unit cells with rectangular holes, $\xi(\mathbf{x})$ may be considered as

$$\xi(\mathbf{x}) = 1 - a(\mathbf{x})b(\mathbf{x}).$$

Assuming isotropy for the material, the matrix of the artificial elasticity constants of a plane stress problem can be written as [11]

$$\mathbf{D}^H = \frac{(1-ab)^\mu E}{(1-\nu^2)} \begin{bmatrix} 1 & \nu & 0 \\ \nu & 1 & 0 \\ 0 & 0 & (1-\nu)/2 \end{bmatrix}. \tag{3.20}$$

Note that $\mathbf{D}^H$ in (3.20) belongs to an isotropic material.

This idea can also be extended to an orthotropic material model which it is believed that it relieves the dependency of results on finite element meshes [18, 23]. The elasticity matrix $\mathbf{D}^H$ can have the form

$$\mathbf{D}^H = \frac{1}{(1-\nu^2)} \begin{bmatrix} E_1 & \nu\sqrt{E_1E_2} & 0 \\ \nu\sqrt{E_1E_2} & E_2 & 0 \\ 0 & 0 & G \end{bmatrix} \tag{3.21a}$$

where $G = (1-\nu)\sqrt{E_1E_2}/2$ and $\xi = \xi_1 + \xi_2 - \xi_1\xi_2$. ξ_1 and ξ_2 are the independent artificial density functions and $E_1 = E^0\xi_1^\mu$, $E_2 = E^0\xi_2^\mu$ and

$\rho = \xi\rho^0$. E^0, ν and ρ^0 are the modulus of elasticity, Poisson's ratio and density of material, respectively [23].

In an analogy to the material model with rectangular voids and inspired by (3.21a), the following orthotropic material model may be constructed:

$$\mathbf{D}^H = \frac{(1-ab)^\gamma E}{(1-\nu^2)}$$

$$\begin{bmatrix} (1-b)^\mu & \nu[(1-a)(1-b)]^{\frac{\mu}{2}} & 0 \\ \nu[(1-a)(1-b)]^{\frac{\mu}{2}} & (1-a)^\mu & 0 \\ 0 & 0 & \frac{(1-\nu)}{2}[(1-a)(1-b)]^{\frac{\mu}{2}} \end{bmatrix} \quad (3.21b)$$

where the exponents γ and μ need to be found by experiments.

There is some controversy about the verification of using the artificial material model in structural topology optimization problems. Bendsøe [9] believes that "the scheme is very dependent on the mesh and it is impossible to give any physical meaning to intermediate values of ξ". Rozvany *et al* has questioned the above statement: "The authors of this paper do not find, however, that the results are highly mesh dependent, nor that a physical interpretation of this model is impossible" [24].

3.3 Analytical solution of the homogenization equations for laminate composites

Recalling from Section 2.6 in the case of lack of the traction forces inside the microcells of the cellular body, which is the case used in topological structural optimization, the governing equations (2.76) and (2.80) are

$$\int_{\mathbf{Y}} E_{ijpq} \frac{\partial \chi_p^{k\ell}}{\partial y_q} \frac{\partial v_i(\mathbf{y})}{\partial y_j} dY = \int_{\mathbf{Y}} E_{ijk\ell} \frac{\partial v_i(\mathbf{y})}{\partial y_j} dY \qquad \forall v \in \mathbf{V_Y}, \quad (3.22)$$

$$E^H_{ijk\ell}(\mathbf{x}) = \frac{1}{|Y|} \int_{\mathbf{Y}} \left(E_{ijk\ell} - E_{ijpq} \frac{\partial \chi_p^{k\ell}}{\partial y_q} \right) dY. \quad (3.23)$$

For the rank laminate composites these equations can be solved analytically. The microscopic characteristic displacement field $\chi^{k\ell}$ is the $\mathbf{Y}$-periodic solution of (3.22) and it is used in (3.23) to find the effective elasticity coefficients. Equations (3.22) and (3.23) with different values of k and ℓ provide enough equations to find the elements of the homogenized matrix $\mathbf{D}^H$. For 2D problems $(i, j, k, \ell, p, q = 1, 2)$ it would be sufficient to solve them for the cases $(\mathbf{a} : k = \ell = 1)$, $(\mathbf{b} : k = \ell = 2)$ and $(\mathbf{c} : k = 1, \ell = 2)$.

case a: $k = 1$, $\ell = 1$

Expanding (3.22) and assuming orthotropicity for the material and removing terms with zero coefficients, we get

$$\int_Y \left[\left(E_{1111} \frac{\partial \chi_1^{11}}{\partial y_1} + E_{1122} \frac{\partial \chi_2^{11}}{\partial y_2} \right) \frac{\partial v_1}{\partial y_1} + E_{1212} \left(\frac{\partial \chi_1^{11}}{\partial y_2} + \frac{\partial \chi_2^{11}}{\partial y_1} \right) \left(\frac{\partial v_1}{\partial y_2} + \frac{\partial v_2}{\partial y_1} \right) + \right.$$

$$\left. \left(E_{1122} \frac{\partial \chi_1^{11}}{\partial y_1} + E_{2222} \frac{\partial \chi_2^{11}}{\partial y_2} \right) \frac{\partial v_2}{\partial y_2} \right] dY = \int_Y \left(E_{1111} \frac{\partial v_1}{\partial y_1} + E_{1122} \frac{\partial v_2}{\partial y_2} \right) dY \tag{3.24}$$

and considering the case $i = j = 1$, from (3.23) it follows that

$$E_{1111}^H = \frac{1}{|Y|} \int_Y \left(E_{1111} - E_{1111} \frac{\partial \chi_1^{11}}{\partial y_1} - E_{1122} \frac{\partial \chi_2^{11}}{\partial y_2} \right) dY \tag{3.25}$$

and if $i = j = 2$ from (3.23) we obtain

$$E_{2211}^H = \frac{1}{|Y|} \int_Y \left(E_{2211} - E_{2211} \frac{\partial \chi_1^{11}}{\partial y_1} - E_{2222} \frac{\partial \chi_2^{11}}{\partial y_2} \right) dY. \tag{3.26}$$

case b: $k = 2$, $\ell = 2$

Following the same procedure as above (3.22) results in

$$\int_Y \left[\left(E_{1111} \frac{\partial \chi_1^{22}}{\partial y_1} + E_{1122} \frac{\partial \chi_2^{22}}{\partial y_2} \right) \frac{\partial v_1}{\partial y_1} + E_{1212} \left(\frac{\partial \chi_1^{22}}{\partial y_2} + \frac{\partial \chi_2^{22}}{\partial y_1} \right) \left(\frac{\partial v_1}{\partial y_2} + \frac{\partial v_2}{\partial y_1} \right) + \right.$$

$$\left. \left(E_{2211} \frac{\partial \chi_1^{22}}{\partial y_1} + E_{2222} \frac{\partial \chi_2^{22}}{\partial y_2} \right) \frac{\partial v_2}{\partial y_2} \right] dY = \int_Y \left(E_{1122} \frac{\partial v_1}{\partial y_1} + E_{2222} \frac{\partial v_2}{\partial y_2} \right) dY. \tag{3.27}$$

If $i = j = 1$, (3.23) becomes

$$E_{1122}^H = \frac{1}{|Y|} \int_Y \left(E_{1122} - E_{1111} \frac{\partial \chi_1^{22}}{\partial y_1} - E_{1122} \frac{\partial \chi_2^{22}}{\partial y_2} \right) dY \tag{3.28}$$

and for $i = j = 2$, (3.23) yields

$$E_{2222}^H = \frac{1}{|Y|} \int_Y \left(E_{2222} - E_{2211} \frac{\partial \chi_1^{22}}{\partial y_1} - E_{2222} \frac{\partial \chi_2^{22}}{\partial y_2} \right) dY. \tag{3.29}$$

case c: $k = 1$, $\ell = 2$

Similarly, in this case from (3.22) we obtain

$$\int_Y \left[\left(E_{1111} \frac{\partial \chi_1^{12}}{\partial y_1} + E_{1122} \frac{\partial \chi_2^{12}}{\partial y_2} \right) \frac{\partial v_1}{\partial y_1} + E_{1212} \left(\frac{\partial \chi_1^{12}}{\partial y_2} + \frac{\partial \chi_2^{12}}{\partial y_1} \right) \left(\frac{\partial v_1}{\partial y_2} + \frac{\partial v_2}{\partial y_1} \right) + \right.$$

$$\left. \left(E_{1122} \frac{\partial \chi_1^{12}}{\partial y_1} + E_{2222} \frac{\partial \chi_2^{12}}{\partial y_2} \right) \frac{\partial v_2}{\partial y_2} \right] dY = \int_Y E_{1212} \left(\frac{\partial v_1}{\partial y_1} + \frac{\partial v_2}{\partial y_2} \right) dY \tag{3.30}$$

and from (3.23) for $i = 1$ and $j = 2$ it follows that

$$E^H_{1212} = \frac{1}{|Y|}\int_Y E_{1212}\left(1 - \frac{\partial \chi_1^{12}}{\partial y_2} - \frac{\partial \chi_2^{12}}{\partial y_1}\right) dY. \tag{3.31}$$

3.3.1 Rank-1 materials

Figure 3.5 illustrates a rank-1 material comprising layers in the y_2 direction and repeated periodically along the y_1 axis. In this case the unit cell $\mathbf{Y} =]0,1[\times I\!R$ consists of two isotropic materials with relative thicknesses γ (strong material) and $1-\gamma$ (weak material). We note that the $\mathbf{Y}$-periodicity assumption implies that $\chi^{k\ell}$ and the test functions v^i are independent of the y_2 coordinate.

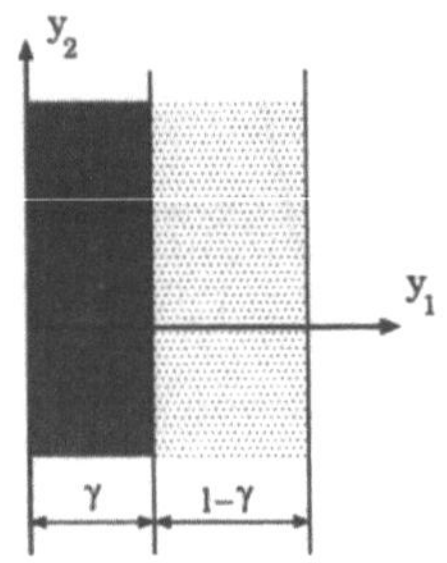

Figure 3.5 *A rank-1 layered material*

case a:

Now considering case (a) and making use of the fact that the functions χ and v are only dependent on y_1 and using test functions $v = \{\varphi(y_1), 0\}^T$, from (3.24) it follows that

$$\int_Y E_{1111}\,\frac{\partial \chi_1^{11}}{\partial y_1}\,\frac{\partial v_1}{\partial y_1} dY = \int_Y E_{1111}\,\frac{\partial v_1}{\partial y_1} dY \tag{3.32}$$

or

$$\int_Y \left(E_{1111}\,\frac{\partial \chi_1^{11}}{\partial y_1} - E_{1111}\right)\frac{\partial v_1}{\partial y_1} dY = 0. \tag{3.33}$$

According to fact (2) in Section 2.3 of the last chapter, as v_1 is $\mathbf{Y}$-periodic

$$\int_Y \frac{\partial v_1}{\partial y_1} dY = 0. \tag{3.34}$$

Comparing (3.33) and (3.34) results in

$$E_{1111}\,\frac{\partial \chi_1^{11}}{\partial y_1} - E_{1111} = C_1 \tag{3.35}$$

where C_1 is a constant. Rearranging (3.35) as

$$\frac{\partial \chi_1^{11}}{\partial y_1} = 1 + \frac{C_1}{E_{1111}}, \tag{3.36}$$

from **Y**-periodicity of χ^{11} and fact (2) we obtain

$$C_1 = -\frac{1}{\dfrac{1}{|Y|}\displaystyle\int_Y \frac{dy}{E_{1111}}} \tag{3.37}$$

If we use the following notation for the volumetric average of a function f over Y,

$$\langle f \rangle = \frac{1}{|Y|}\int_Y f(y)\,dy, \tag{3.38}$$

then (3.37) can be written as

$$C_1 = -\left\langle \frac{1}{E_{1111}} \right\rangle^{-1}. \tag{3.39}$$

Also, (3.25) yields

$$E^H_{1111} = \frac{1}{|Y|}\int_Y \left(E_{1111} - E_{1111}\frac{\partial \chi_1^{11}}{\partial y_1} \right) dY = -C_1, \tag{3.40}$$

and consequently

$$E^H_{1111} = \left\langle \frac{1}{E_{1111}} \right\rangle^{-1}, \tag{3.41}$$

or by the compact notation (see Appendix A)

$$\boxed{D^H_{11} = \left\langle D^{-1}_{11} \right\rangle^{-1}} \; . \tag{3.42}$$

case b:

Similarly, in case (b) by assuming the same test functions as before, from (3.27) we obtain

$$\int_Y \left(E_{1111}\frac{\partial \chi_1^{22}}{\partial y_1} - E_{1122} \right) \frac{\partial v_1}{\partial y_1} dY = 0 \tag{3.43}$$

which gives

$$E_{1111}\frac{\partial \chi_1^{22}}{\partial y_1} - E_{1122} = C_2, \tag{3.44}$$

or

$$\frac{\partial \chi_1^{22}}{\partial y_1} = \frac{E_{1122}}{E_{1111}} + \frac{C_2}{E_{1111}} \; . \tag{3.45}$$

Integrating this equation on Y and using fact (2) results in

$$\frac{1}{|Y|}\int_Y\left(\frac{E_{1122}}{E_{1111}}+\frac{C_2}{E_{1111}}\right)dY=0 \tag{3.46}$$

or

$$C_2\left\langle\frac{1}{E_{1111}}\right\rangle+\left\langle\frac{E_{1122}}{E_{1111}}\right\rangle=0\,. \tag{3.47}$$

Thus

$$C_2=-\left\langle\frac{E_{1122}}{E_{1111}}\right\rangle\left\langle\frac{1}{E_{1111}}\right\rangle^{-1}. \tag{3.48}$$

On the other hand (3.28) yields

$$E^H_{1122}=\frac{1}{|Y|}\int_Y\left(E_{1122}-E_{1111}\frac{\partial\chi_1^{22}}{\partial y_1}\right)dY. \tag{3.49}$$

Comparing (3.49) and (3.44) one obtains

$$E^H_{1122}=-C_2 \tag{3.50}$$

and by using the compact notation we conclude that

$$\boxed{D^H_{12}=\left\langle D_{12}D_{11}^{-1}\right\rangle\left\langle D_{11}^{-1}\right\rangle^{-1}}\,. \tag{3.51}$$

Now since we know D^H_{11} and D^H_{12} from (3.29), it can be shown that

$$E^H_{2222}=\frac{1}{|Y|}\int_Y\left(E_{2222}-E_{2211}\frac{\partial\chi_1^{22}}{\partial y_1}\right)dY \tag{3.52}$$

and by introducing (3.45) and (3.48) into (3.52) and using the compact notation we obtain

$$\boxed{D^H_{22}=\langle D_{22}\rangle-\left\langle D_{12}^2D_{11}^{-1}\right\rangle+\left\langle D_{12}D_{11}^{-1}\right\rangle^2\left\langle D_{11}^{-1}\right\rangle^{-1}}\,. \tag{3.53}$$

case c:

Finally, considering case (c) where $k=1$ and $\ell=2$, and using $v=[0,\varphi(y_1)]^T$ as the test function, (3.30) yields

$$\int_Y\left(E_{1212}\frac{\partial\chi_2^{12}}{\partial y_1}-E_{1212}\right)\frac{\partial v_2}{\partial y_1}dY=0 \tag{3.54}$$

which following the same procedure as before results in

$$E_{1212}\frac{\partial\chi_2^{12}}{\partial y_1}-E_{1212}=C_3 \tag{3.55}$$

and as for the last two cases it is concluded that

$$C_3 = -\left\langle \frac{1}{E_{1212}} \right\rangle^{-1} . \tag{3.56}$$

From (3.31) it follows that

$$E^H_{1212} = \frac{1}{|Y|} \int_Y E_{1212} \left(1 - \frac{\partial \chi_2^{12}}{\partial y_1} \right) dY = -C_3 . \tag{3.57}$$

Therefore from (2.56) and (2.57) and using the compact notation (see Appendix A) we obtain

$$\boxed{D^H_{66} = \left\langle D^{-1}_{66} \right\rangle^{-1}} . \tag{3.58}$$

For the unit cell of Figure 3.5, consisting of material I (strong) and material II (weak) with thicknesses γ and $1-\gamma$ respectively, from (3.42) it follows that

$$D^H_{11} = \left\langle D^{-1}_{11} \right\rangle^{-1} = 1/\left[\int_0^\gamma \frac{dy_1}{D^I_{11}} + \int_\gamma^1 \frac{dy_1}{D^{II}_{11}} \right] \tag{3.59}$$

$$D^H_{11} = \frac{D^I_{11} D^{II}_{11}}{\gamma D^{II}_{11} + (1-\gamma) D^I_{11}} . \tag{3.60}$$

Now let us define the following terms:

Arithmetic average $\overline{A}$. For all $\xi \in [0,1]$, the arithmetic average of two real numbers r_1 and r_2 is defined as

$$\overline{A}(r_1, r_2, \xi) = \xi r_1 + (1-\xi) r_2 . \tag{3.61}$$

Harmonic average $\underline{A}$. For all $\xi \in [0,1]$, the harmonic average of two real numbers r_1 and r_2 is defined as

$$\underline{A}(r_1, r_2, \xi) = \frac{r_1 r_2}{\xi r_2 + (1-\xi) r_1} . \tag{3.62}$$

For positive values of r_1 and r_2, $\underline{A} \leq \overline{A}$ and $\underline{A} = \overline{A}$ if and only if $r_1 = r_2$.

By making use of the above definitions, (3.60) can be written as

$$D^H_{11} = \underline{A}\left(D^I_{11}, D^{II}_{11}, \gamma \right) . \tag{3.63}$$

Similarly, from (3.51), (3.53) and (3.58) respectively, it follows that

$$D^H_{12} = \overline{A}\left({D_{12}}^I {D^I_{11}}^{-1}, {D_{12}}^{II} {D^{II}_{11}}^{-1}, \gamma \right) \underline{A}\left(D^I_{11}, D^{II}_{11}, \gamma \right) , \tag{3.64}$$

$$D_{22}^{H} = \overline{A}\left(D_{22}^{I}, D_{22}^{II}, \gamma\right) - \overline{A}\left({D_{12}^{I}}^{2} {D_{11}^{I}}^{-1}, {D_{12}^{II}}^{2} {D_{11}^{II}}^{-1}, \gamma\right)$$
$$+\overline{A}^{2}\left({D_{12}}^{I} {D_{11}^{I}}^{-1}, {D_{12}}^{II} {D_{11}^{II}}^{-1}, \gamma\right) \underline{A}\left(D_{11}^{I}, D_{11}^{II}, \gamma\right) \tag{3.65}$$

and

$$D_{66}^{H} = \underline{A}\left(D_{66}^{I}, D_{66}^{II}, \gamma\right). \tag{3.66}$$

If the materials are isotropic with the same Poisson's ratio and different Young's moduli E^{I} and E^{II}, then

$$D_{11}^{i} = D_{22}^{i} = \frac{E^{i}}{1-\nu^{2}} \quad ; \quad D_{12}^{i} = \frac{\nu}{1-\nu^{2}} E^{i} \quad ; \quad D_{66}^{i} = \frac{E^{i}}{2(1+\nu)} \qquad (i = I, II).$$

In this case by using (3.63)-(3.66) one obtains

$$\mathbf{D}^{H} = \begin{bmatrix} J_1 & \nu J_1 & 0 \\ \nu J_1 & J_2 + \nu^{2} J_1 & 0 \\ 0 & 0 & (1-\nu)/2 J_1 \end{bmatrix}. \tag{3.67}$$

where

$$J_1 = \frac{1}{1-\nu^{2}} \underline{A}\left(E^{I}, E^{II}, \gamma\right) \quad ; \quad J_2 = \overline{A}\left(E^{I}, E^{II}, \gamma\right).$$

In the particular case when the weak material is complete void (i.e. $E_{II} = 0$), we have $J_1 = 0$ and $J_2 = \gamma E^{I}$. Thus, (3.67) becomes

$$\mathbf{D}^{H} = \begin{bmatrix} 0 & 0 & 0 \\ 0 & \gamma E^{I} & 0 \\ 0 & 0 & 0 \end{bmatrix}. \tag{3.68}$$

3.3.2 Rank-2 materials

As explained in Section 3.2.2, the rank-1 material can be used as one of the components of a rank-2 material (see Figure 3.3). Having relative thicknesses μ for the strong material and $1-\mu$ for the already-constructed rank-1 material, as shown in Figure 3.6, by recursive use of (3.63)-(3.66) the elements of the elasticity matrix can easily be found.

For example, assuming isotropy for the solid part of the material with the Young's modulus E and Poisson's ratio ν and noticing that the directions of the rank-1 layers and the solid layers are perpendicular, from (3.63) it follows that

$$D_{11}^{H} = \underline{A}\left(D_{11}^{I}, D_{22}^{R1}, \mu\right) \tag{3.69}$$

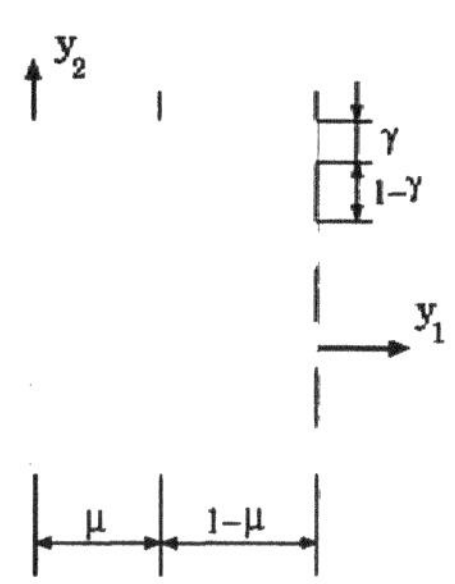

Figure 3.6 Rank-2 material

where $D_{11}^{I} = E/(1-\nu^2)$ and $D_{22}^{R1} = \gamma E$. Here $R1$ stands for the rank-1 material, also see (3.68). Therefore, using (3.62) we get

$$D_{11}^{H} = \frac{\gamma E}{(1-\mu) + \gamma\mu(1-\nu^2)} \tag{3.70}$$

Similarly, the other elements of the homogenized matrix can easily be found,

$$\boxed{\mathbf{D}^{H} = \frac{E}{(1-\mu) + \gamma\mu(1-\nu^2)} \begin{bmatrix} \gamma & \mu\gamma\nu & 0 \\ \mu\gamma\nu & \mu(1-\mu+\mu\gamma) & 0 \\ 0 & 0 & 0 \end{bmatrix}} . \tag{3.71}$$

In practice in order to avoid singularity in the optimization algorithm, the voids are replaced with a very weak material. The contribution of this material to the homogenized matrix can easily be calculated using the procedure described earlier.

3.3.3 Bi-material rank-2 composites

Referring to Figure 3.4 and the instructions for the construction of the bi-material rank-2 composites and using the superscripts I, II, V, III and $R1$ for the stiffer solid, the less stiff solid, the very weak material which represents the voids, the composite comprising two solids (Figure 3.4(a)) and the rank-1 bi-material respectively, by recursively recalling (3.63)-(3.66) and noticing that the axes of the rank-1 bi-material and the material I are perpendicular to each other, we obtain

$$\boxed{D_{11}^{H} = \underline{A}\left(D_{11}^{I}, D_{22}^{R1}, \mu\right)} \tag{3.72}$$

where

$$\begin{aligned} D_{22}^{R1} = {} & \overline{A}\left(D_{22}^{III}, D_{22}^{V}, \gamma_2\right) - \overline{A}\left({D_{12}^{III}}^2 {D_{11}^{III}}^{-1}, {D_{12}^{V}}^2 {D_{11}^{V}}^{-1}, \gamma_2\right) \\ & + \overline{A}^2\left({D_{12}}^{III} {D_{11}^{III}}^{-1}, {D_{12}}^{V} {D_{11}^{V}}^{-1}, \gamma_2\right) \underline{A}\left({D_{11}}^{III}, {D_{11}}^{V}, \gamma_2\right) \end{aligned} \tag{3.73}$$

and $D_{11}^{III}, \cdots, D_{66}^{III}$ are directly derived from (3.63)-(3.66). For example

$$D_{11}^{III} = \underline{A}\left(D_{11}^{I}, D_{11}^{II}, \gamma_1\right) \quad , \cdots$$

Similarly,

$$\boxed{D_{12}^{H} = \overline{A}\left({D_{12}}^{I} {D_{11}^{I}}^{-1}, {D_{12}}^{R1} {D_{22}^{R1}}^{-1}, \mu\right) \underline{A}\left(D_{11}^{I}, D_{22}^{II}, \mu\right)} \quad , \tag{3.74}$$

$$\boxed{\begin{aligned} D_{22}^{H} = {} & \overline{A}\left(D_{22}^{I}, D_{11}^{R1}, \mu\right) - \overline{A}\left({D_{12}^{I}}^{2} {D_{11}^{I}}^{-1}, {D_{12}^{R1}}^{2} {D_{22}^{R1}}^{-1}, \mu\right) \\ & + \overline{A}^{2}\left({D_{12}}^{I} {D_{11}^{I}}^{-1}, {D_{12}}^{R1} {D_{22}^{R1}}^{-1}, \mu\right) \underline{A}\left(D_{11}^{I}, D_{22}^{R1}, \mu\right) \end{aligned}} \quad . \tag{3.75}$$

Finally,

$$\boxed{D_{66}^{H} = \underline{A}\left(D_{66}^{I}, D_{66}^{R1}, \mu\right)} \quad . \tag{3.76}$$

3.4 Numerical solution of the homogenization equations for a cellular body with rectangular holes

For most geometries of the base cell, the homogenization equations have to be solved by numerical techniques such as the finite element, boundary element or spectral methods. In the next section, the derivation of the finite element formulation for two dimensional problems is explained.

3.4.1 Finite element formulation

Following a procedure similar to the analytical solution, the same three cases are considered.

case a: $k = 1 \ , \ \ell = 1$

For the sake of simplicity by using the compact notation (see Appendix A) and $\chi_1^{11} = \Phi_1$ and $\chi_2^{11} = \Phi_2$, (3.24) and (3.25) will become

$$\begin{aligned} & \int_Y \left[\left(D_{11} \frac{\partial \Phi_1}{\partial y_1} + D_{12} \frac{\partial \Phi_2}{\partial y_2}\right) \frac{\partial v_1}{\partial y_1} + D_{66} \left(\frac{\partial \Phi_1}{\partial y_2} + \frac{\partial \Phi_2}{\partial y_1}\right) \left(\frac{\partial v_1}{\partial y_2} + \frac{\partial v_2}{\partial y_1}\right) \right. \\ & \left. + \left(D_{12} \frac{\partial \Phi_1}{\partial y_1} + D_{22} \frac{\partial \Phi_2}{\partial y_2}\right) \frac{\partial v_2}{\partial y_2}\right] dY = \int_Y \left(D_{11} \frac{\partial v_1}{\partial y_1} + D_{12} \frac{\partial v_2}{\partial y_2}\right) dY \end{aligned} \tag{3.77}$$

and

$$D_{11}^{H} = \frac{1}{|Y|} \int_Y \left(D_{11} - D_{11} \frac{\partial \Phi_1}{\partial y_1} - D_{12} \frac{\partial \Phi_2}{\partial y_2}\right) dY. \tag{3.78}$$

Rearranging (3.77) it follows that

$$\int_Y \left\{ \frac{\partial v_1}{\partial y_1} \quad \frac{\partial v_2}{\partial y_2} \quad \frac{\partial v_1}{\partial y_2} + \frac{\partial v_2}{\partial y_1} \right\} \begin{bmatrix} D_{11} & D_{12} & 0 \\ D_{12} & D_{22} & 0 \\ 0 & 0 & D_{66} \end{bmatrix} \left\{ \begin{array}{c} \partial\Phi_1/\partial y_1 \\ \partial\Phi_2/\partial y_2 \\ \partial\Phi_1/\partial y_2 + \partial\Phi_2/\partial y_1 \end{array} \right\} dY$$

$$= \int_Y \left\{ \frac{\partial v_1}{\partial y_1} \quad \frac{\partial v_2}{\partial y_2} \quad \frac{\partial v_1}{\partial y_2} + \frac{\partial v_2}{\partial y_1} \right\} \left\{ \begin{array}{c} D_{11} \\ D_{12} \\ 0 \end{array} \right\} dY. \tag{3.79}$$

Now let us define

$$\epsilon(v) = \left\{ \begin{array}{c} \partial v_1/\partial y_1 \\ \partial v_2/\partial y_2 \\ \partial v_1/\partial y_2 + \partial v_2/\partial y_1 \end{array} \right\}, \quad \epsilon(\mathbf{\Phi}) = \left\{ \begin{array}{c} \partial\Phi_1/\partial y_1 \\ \partial\Phi_2/\partial y_2 \\ \partial\Phi_1/\partial y_2 + \partial\Phi_2/\partial y_1 \end{array} \right\} \tag{3.80}$$

and

$$\mathbf{D} = [\, \mathbf{d}_1 \quad \mathbf{d}_2 \quad \mathbf{d}_3 \,] \tag{3.81}$$

where $\mathbf{d}_1$, $\mathbf{d}_2$ and $\mathbf{d}_3$ are the columns of the elasticity matrix $\mathbf{D}$, so that

$$\mathbf{d}_1 = \left\{ \begin{array}{c} D_{11} \\ D_{12} \\ 0 \end{array} \right\}, \quad \mathbf{d}_2 = \left\{ \begin{array}{c} D_{12} \\ D_{22} \\ 0 \end{array} \right\} \text{ and } \mathbf{d}_3 = \left\{ \begin{array}{c} 0 \\ 0 \\ D_{66} \end{array} \right\}. \tag{3.82}$$

Using the above definitions (3.79) can be written as

$$\int_Y \epsilon^T(v)\, \mathbf{D}\, \epsilon(\mathbf{\Phi})\, dY = \int_Y \epsilon^T(v)\, \mathbf{d}_1\, dY \qquad \forall v \in \mathbf{V_Y} \tag{3.83}$$

and similarly (3.78) becomes

$$\boxed{D_{11}^H = \frac{1}{|Y|} \int_Y \left(D_{11} - \mathbf{d}_1^T\, \epsilon(\mathbf{\Phi}) \right) dY} \;. \tag{3.84}$$

Now we can apply the usual finite element method [25-28]. Discretizing the domain of the base cell and using global shape functions we can write

$$v = \sum_{i=1}^{np} \mathbf{N}_i^g\, \hat{v}_i = \mathbf{N}^g \hat{v} \tag{3.85a}$$

$$\mathbf{\Phi} = \sum_{i=1}^{np} \mathbf{N}_i^g\, \hat{\mathbf{\Phi}}_i = \mathbf{N}^g \hat{\mathbf{\Phi}} \tag{3.85b}$$

where

$$v = [v_1, v_2]^T \quad , \quad \mathbf{\Phi} = [\Phi_1, \Phi_2]^T, \tag{3.86a}$$

$$\hat{v}_i = [v_{1i}, v_{2i}]^T \quad , \quad \hat{\mathbf{\Phi}}_i = [\Phi_{1i}, \Phi_{2i}]^T, \tag{3.86b}$$

$$\hat{v} = [\hat{v}_1, \hat{v}_2, \cdots, \hat{v}_{np}]^T \quad , \quad \hat{\mathbf{\Phi}} = [\hat{\mathbf{\Phi}}_1, \hat{\mathbf{\Phi}}_2, \cdots, \hat{\mathbf{\Phi}}_{np}]^T \tag{3.86c}$$

and

$$\mathbf{N}_i^g = \begin{bmatrix} 1 & 0 \\ 0 & 1 \end{bmatrix} N_i^g \tag{3.87}$$

where N_i^g is the global shape function associated with node i and np is the number of nodes in the whole discretized domain. Applying this approximation to (3.80) results in

$$\epsilon(\boldsymbol{v}) = \mathbf{L}\boldsymbol{v} = \mathbf{L}\mathbf{N}^g\hat{\boldsymbol{v}} = \mathbf{B}\hat{\boldsymbol{v}} \tag{3.88a}$$

$$\epsilon(\boldsymbol{\Phi}) = \mathbf{L}\boldsymbol{\Phi} = \mathbf{L}\mathbf{N}^g\hat{\boldsymbol{\Phi}} = \mathbf{B}\hat{\boldsymbol{\Phi}} \tag{3.88b}$$

where $\mathbf{L}$ is the matrix of linear differential operators defined as

$$\mathbf{L} = \begin{bmatrix} \partial/\partial y_1 & 0 \\ 0 & \partial/\partial y_2 \\ \partial/\partial y_2 & \partial/\partial y_1 \end{bmatrix} \tag{3.89}$$

and

$$\mathbf{B} = \mathbf{L}\mathbf{N}^g \tag{3.90}$$

is the global strain matrix.

Now substituting (3.89) and (3.90) into (3.83) we obtain

$$\hat{\boldsymbol{v}}^T \int_Y \mathbf{B}^T\mathbf{D}\,\mathbf{B}\,dY\,\hat{\boldsymbol{\Phi}} = \hat{\boldsymbol{v}}^T \int_Y \mathbf{B}^T\mathbf{d}_1\,dY \qquad \forall \boldsymbol{v}. \tag{3.91}$$

As $\boldsymbol{v}$ is an arbitrary test function, by removing $\hat{\boldsymbol{v}}$ from both sides of (3.91) we get

$$\int_Y \mathbf{B}^T\mathbf{D}\,\mathbf{B}\,dY\,\hat{\boldsymbol{\Phi}} = \int_Y \mathbf{B}^T\mathbf{d}_1\,dY \tag{3.92}$$

which is very similar to the well known stiffness equation

$$\boxed{\mathbf{K}\,\hat{\boldsymbol{\Phi}} = \mathbf{f}}\ , \tag{3.93}$$

where the stiffness matrix is written as

$$\mathbf{K} = \int_Y \mathbf{B}^T\mathbf{D}\mathbf{B}\,dY \tag{3.94}$$

and the 'force' vector has the form

$$\mathbf{f} = \int_Y \mathbf{B}^T\mathbf{d}_1\,dY. \tag{3.95}$$

If n-node elements are used to discretize the domain, then within a typical element e

$$\boldsymbol{\Phi} = \sum_{i=1}^{n} \mathbf{N}_i^e\,\hat{\boldsymbol{\Phi}}_i \tag{3.96}$$

where

$$\mathbf{N}_i^e = \begin{bmatrix} 1 & 0 \\ 0 & 1 \end{bmatrix} N_i^e \quad \text{and} \quad \hat{\boldsymbol{\Phi}}_i = \begin{Bmatrix} \hat{\Phi}_{1i} \\ \hat{\Phi}_{2i} \end{Bmatrix}. \tag{3.97}$$

In (3.96) and (3.97) N_i^e is the shape function associated with node i of the element e and n is the number of nodes of each element. From (3.96) it follows that

$$\epsilon(\mathbf{\Phi}) = \sum_{i=1}^{n} \mathbf{B}_i^e \hat{\mathbf{\Phi}}_i \tag{3.98}$$

where

$$\mathbf{B}_i^e = \begin{bmatrix} \partial N_i^e/\partial y_1 & 0 \\ 0 & \partial N_i^e/\partial y_2 \\ \partial N_i^e/\partial y_2 & \partial N_i^e/\partial y_1 \end{bmatrix}. \tag{3.99}$$

Expression (3.99) enables us to find the submatrix of the stiffness matrix of element e linking nodes i and j

$$\boxed{\mathbf{k}_{ij}^e = \int_{Y^e} \mathbf{B}_i^{eT} \mathbf{D} \mathbf{B}_j^e \, dY} \tag{3.100}$$

where Y^e denotes the volume of element. Also the nodal forces of element e associated with node i are obtained from

$$\boxed{\mathbf{f}_i^e = \int_{Y^e} \mathbf{B}_i^{eT} \mathbf{d}_1 dY}. \tag{3.101}$$

We also note that the global stiffness matrix and the global forces can be assembled from the element stiffness matrices and force vectors

$$\mathbf{K} = \mathop{\mathbf{A}}_{e=1}^{m} \mathbf{k}^e \tag{3.102}$$

$$\mathbf{f} = \mathop{\mathbf{A}}_{e=1}^{m} \mathbf{f}^e \tag{3.103}$$

where $\mathop{\mathbf{A}}_{e=1}^{m}$ stands for the finite element assembly operator and m is the number of elements.

It is interesting to notice that the 'force' vector used in homogenization has a physical meaning and it is, in fact, a specific case of *initial strain loading.* To demonstrate this we recall that the nodal forces induced by initial strains are (References [26] and [28])

$$\{\mathbf{f}_i^{\epsilon^0}\}^e = \int_{Y^e} \mathbf{B}^{eT} \mathbf{D} \epsilon^0 \, dY. \tag{3.104}$$

Now comparing (3.104) and (3.101) results in

$$\mathbf{D}\,\boldsymbol{\epsilon}^0 = \mathbf{d}_1 \tag{3.105}$$

or

$$\begin{bmatrix} D_{11} & D_{12} & 0 \\ D_{12} & D_{22} & 0 \\ 0 & 0 & D_{66} \end{bmatrix} \begin{Bmatrix} \epsilon_{11}^0 \\ \epsilon_{22}^0 \\ 2\epsilon_{12}^0 \end{Bmatrix} = \begin{Bmatrix} D_{11} \\ D_{12} \\ 0 \end{Bmatrix} \tag{3.106}$$

which implies that

$$\boxed{\epsilon_{11}^0 = 1\ ,\ \epsilon_{22}^0 = 0 \ \text{ and } \ \epsilon_{12}^0 = 0} \ . \tag{3.107}$$

Thus, in this case, the loading is a unit initial strain in only the y_1 direction. If (3.94) is solved to give the displacements $\hat{\mathbf{\Phi}}$ and strains $\boldsymbol{\epsilon}(\hat{\mathbf{\Phi}})$, then using (3.84) the first element of the homogenized matrix D_{11}^H can be calculated.

case b: $k = 2\ ,\ \ell = 2$

Starting from (3.27) and following the same procedure as for case (a), and representing χ_1^{22} as Ψ_1 and χ_2^{22} as Ψ_2, one obtains

$$\int_Y \boldsymbol{\epsilon}^T(\boldsymbol{v})\ \mathbf{D}\ \boldsymbol{\epsilon}(\mathbf{\Psi})\, dY = \int_Y \boldsymbol{\epsilon}^T(\boldsymbol{v})\, \mathbf{d}_2\, dY, \qquad \forall \boldsymbol{v}. \tag{3.108}$$

Then, using the same steps as for case (a), we obtain

$$\boxed{\mathbf{K}\,\hat{\mathbf{\Psi}} = \mathbf{f}} \ . \tag{3.109}$$

In this case the formulae for the stiffness matrix are the same as for case (a) and the only difference is in the loading. The nodal forces become

$$\boxed{\mathbf{f}_i^e = \int_{Y^e} \mathbf{B}_i^{eT} \mathbf{d}_2\, dY} \tag{3.110}$$

which implies a unit initial strain in the y_2 direction, so that

$$\boxed{\epsilon_{11}^0 = 0\ ,\ \epsilon_{22}^0 = 1 \ \text{ and } \ \epsilon_{12}^0 = 0} \ . \tag{3.111}$$

On the other hand following a similar process, (3.28) and (3.29) can be written as

$$\boxed{D_{12}^H = \frac{1}{|Y|} \int_Y \left(D_{12} - \mathbf{d}_1^T\, \boldsymbol{\epsilon}(\mathbf{\Psi})\right)\, dY} \ , \tag{3.112}$$

$$\boxed{D_{22}^H = \frac{1}{|Y|} \int_Y \left(D_{22} - \mathbf{d}_2^T\, \boldsymbol{\epsilon}(\mathbf{\Psi})\right)\, dY} \ . \tag{3.113}$$

Having obtained the displacement function $\hat{\mathbf{\Psi}}$ and the strain $\boldsymbol{\epsilon}(\hat{\mathbf{\Psi}})$ by solving (3.109), from (3.112) and (3.113) D_{12}^H and D_{22}^H can be obtained.

case c: $k = 1$, $\ell = 2$

As for cases (a) and (b) we use the compact notation and rename χ_1^{12} and χ_2^{12} respectively as θ_1 and θ_2. From (3.30) it follows that

$$\int_Y \epsilon^T(v)\, \mathbf{D}\, \epsilon(\boldsymbol{\theta})\, dY = \int_Y \epsilon^T(v)\, \mathbf{d}_3\, dY \qquad \forall v. \tag{3.114}$$

The nodal forces of elements in this case are

$$\boxed{\mathbf{f}_i^e = \int_{Y^e} \mathbf{B}_i^{eT} \mathbf{d}_3 dY} \tag{3.115}$$

which implies a unit initial shear strain loading

$$\boxed{\epsilon_{11}^0 = 0\ ,\ \epsilon_{22}^0 = 0 \ \text{ and }\ 2\epsilon_{12}^0 = 1}\,. \tag{3.116}$$

Also applying the above notation, (3.31) yields

$$\boxed{D_{66}^H = \frac{1}{|Y|} \int_Y \left(D_{66} - \mathbf{d}_3^T\, \epsilon(\boldsymbol{\theta})\right)\, dY}\,. \tag{3.117}$$

Thus far we have seen how the finite element formulation for the calculation of the elements of the homogenized elasticity matrix is derived. Furthermore, we have observed that for two dimensional problems, by considering three loading cases it is possible to find all of the elasticity coefficients. In practice, after discretizing the domain of the base cell, it is sufficient to run the finite element program for different unit initial strain cases. Derivation of the required boundary conditions will be discussed in the following section.

3.4.2 Derivation of the boundary conditions from periodicity

So far we have concluded that the integral equation of homogenization, (3.22), is a boundary value problem and it can be reduced, after appropriate discretization, to the conventional form

$$\mathbf{K}\boldsymbol{\chi} = \mathbf{f}, \tag{3.118}$$

which has to be solved on the domain of the unit cell. Any boundary value problem to be solved, needs some information about the initial or boundary conditions. To determine the appropriate boundary conditions we reconsider the main assumption of homogenization theory, i.e. periodicity. The microscopic displacement field $\boldsymbol{\chi}$ is the $\mathbf{Y}$-periodic solution of (3.118). The base cell $\mathbf{Y}$ is illustrated in Figure 3.7.

From the definition of periodicity (2.1), it follows that

$$\boldsymbol{\chi}(y_1, y_2) = \boldsymbol{\chi}(y_1 + Y_1, y_2) = \boldsymbol{\chi}(y_1 + Y_1, y_2 + Y_2) = \boldsymbol{\chi}(y_1, y_2 + Y_2). \tag{3.119}$$

Considering $y_1 = y_1^0$ and $y_2 = y_2^0$, we can write that

$$\boldsymbol{\chi}(y_1^0, y_2^0) = \boldsymbol{\chi}(y_1^0 + Y_1, y_2^0) = \boldsymbol{\chi}(y_1^0 + Y_1, y_2^0 + Y_2) = \boldsymbol{\chi}(y_1^0, y_2^0 + Y_2), \tag{3.120}$$

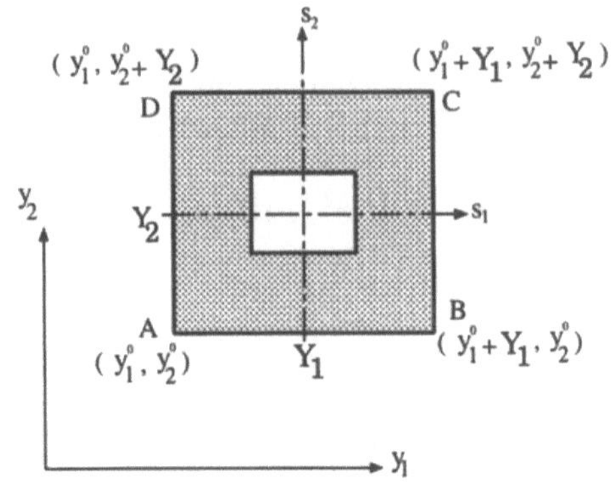

Figure 3.7 *The base cell of periodicity*

which states that the microscopic displacements of the corner points of the base cell are equal ($\overrightarrow{AA'} = \overrightarrow{BB'} = \overrightarrow{CC'} = \overrightarrow{DD'}$). A typical possible deformation of the base cell is exaggerated in Figure 3.8. We notice that this displacement field can be resolved into two parts. One translation or rigid body motion (see Figure 3.9(b)), which has no effect on the stress and strain fields inside the domain of the cell and one in situ deformation (Figure 3.9(c)). In Figure 3.10 a typical microscopic deformation for a group of $\mathbf{Y}$-periodic cells is schematically depicted.

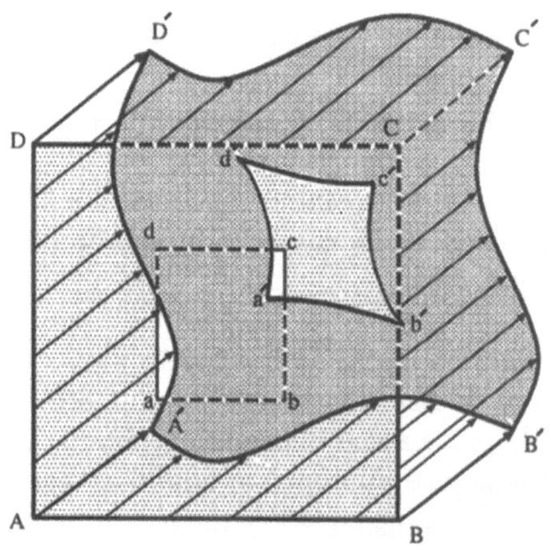

Figure 3.8 *A typical deformation of the base cell*

If the components of χ in the y_1 and y_2 directions, respectively are denoted by the scalar displacement functions u and v, (3.119) can be written as

$$u(y_1, y_2) = u(y_1 + Y_1, y_2) = u(y_1 + Y_1, y_2 + Y_2) = u(y_1, y_2 + Y_2), \quad (3.121)$$

$$v(y_1, y_2) = v(y_1 + Y_1, y_2) = v(y_1 + Y_1, y_2 + Y_2) = v(y_1, y_2 + Y_2). \quad (3.122)$$

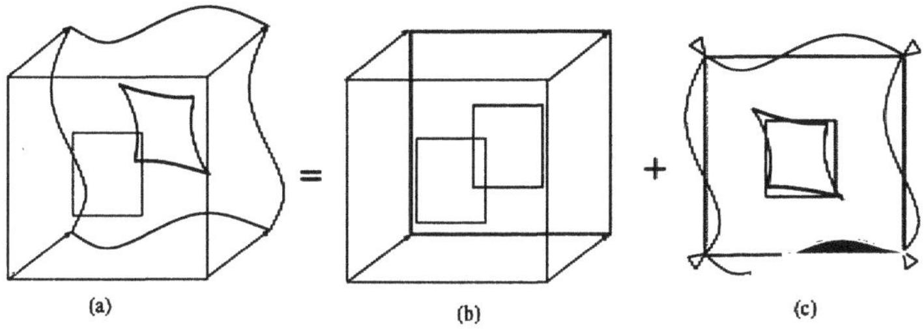

Figure 3.9 *Components of a possible microscopic displacement field*

So far our discussion has been general and (3.120) is valid for any periodic cell of composite material. To obtain the proper boundary restrictions one can utilize the methods applied to the repeatable structures [29]. For example, the numbers of pairs of nodes located on the opposite edges of the cell can be linked so that opposite edges have identical deformed shapes. Alternatively, the Lagrange multiplier method [30] or the penalty method [31, 30] can be adopted. Here a direct method is presented [32, 33].

We focus on cells with rectangular voids which is the material model chosen for the topology optimization†. We remember that for 2D problems the homogenization equation must be solved three times. The earlier cases are again considered here.

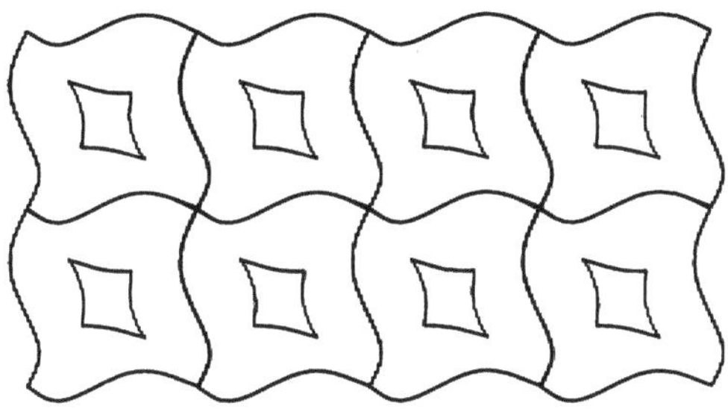

Figure 3.10 *A typical deformation of the cellular body*

case a:

In unit cells with rectangular voids the geometry of the cell is symmetric with respect to the axes of symmetry s_1 and s_2 as illustrated in Figure 3.7. Recalling (3.107), the loading to be imposed in this case is a unit initial strain in the

† However, the following conclusions are valid for all periodic cells of composites which have two orthogonal axes of symmetry.

y_1 direction (i.e. $\epsilon^0_{11} = 1$, $\epsilon^0_{22} = 0$ and $\epsilon^0_{12} = 0$). In Figure 3.11(b) the deformation of the base cell under this loading is depicted. To prevent rigid body motion, the horizontal and vertical displacements of the bottom left corner and the vertical displacement of the bottom right corner are restricted. For the displacements and the applied loading under consideration, we see that the symmetric condition is realized.

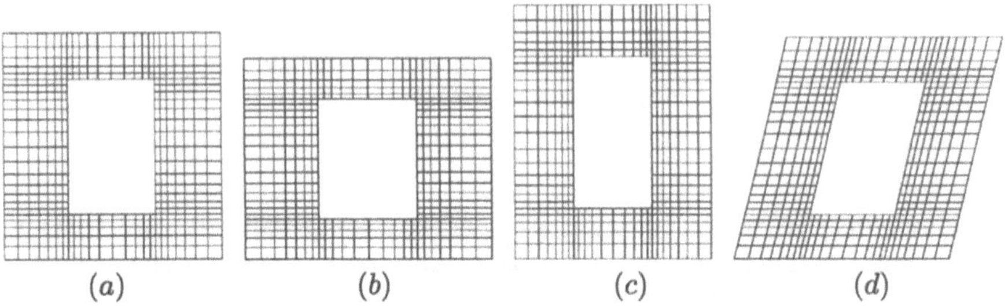

Figure 3.11 (a): Base cell. (b): Deformation induced by unit initial strain in horizontal direction. (c): Deformation induced by unit initial strain in vertical direction. (d): Deformation induced by unit initial shear strain.

From symmetry of the geometry and loading with respect to s_2 (see Figure 3.7) it follows that

$$u(y_1^0, y_2) = -u(y_1^0 + Y_1, y_2), \tag{3.123}$$

$$v(y_1^0, y_2) = v(y_1^0 + Y_1, y_2) \tag{3.124}$$

and similarly from symmetry with respect to s_1 it follows that

$$u(y_1, y_2^0) = u(y_1, y_2^0 + Y_2), \tag{3.125}$$

$$v(y_1, y_2^0) = -v(y_1, y_2^0 + Y_2)\,. \tag{3.126}$$

Comparing (3.123) and (3.126) with (3.121) and (3.122) results in

$$u(y_1^0, y_2) = u(y_1^0 + Y_1, y_2) = 0\,, \tag{3.127}$$

and

$$v(y_1, y_2^0) = v(y_1, y_2^0 + Y_2) = 0\,. \tag{3.128}$$

According to (3.127) and (3.128) the horizontal displacement of the left and right hand side edges of the cell is restricted. Similarly, in the bottom and top edges of the cell the vertical displacement is zero. These boundary conditions

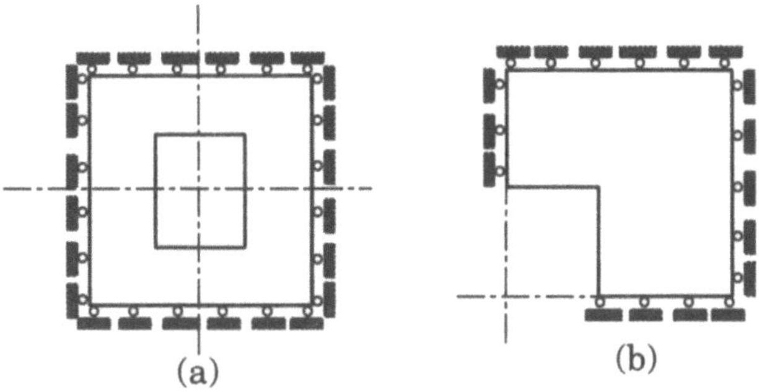

Figure 3.12 *Boundary conditions for cases* **a** *and* **b**

are depicted in Figure 3.12(a). We note that by utilizing symmetry only a quarter of the domain of the unit cell needs to be analyzed; see Figure 3.12(b). We also note that in this case from (3.127) and (3.121), by substituting $y_1 = y_1^0$ and $y_2 = y_2^0$ it follows that

$$u(y_1^0, y_2^0) = u(y_1^0 + Y_1, y_2^0) = u(y_1^0 + Y_1, y_2^0 + Y_2) = u(y_1^0, y_2^0 + Y_2) = 0 \quad (3.129)$$

and similarly (3.128) and (3.122) results in

$$v(y_1^0, y_2^0) = v(y_1^0 + Y_1, y_2^0) = v(y_1^0 + Y_1, y_2^0 + Y_2) = v(y_1^0, y_2^0 + Y_2) = 0 \quad (3.130)$$

which means that the rigid body motion does not happen. The deformation of the base cell is depicted in Figure 3.13. As the deformations induced by unit initial strain are very large, they are scaled in the ratio 1:4.

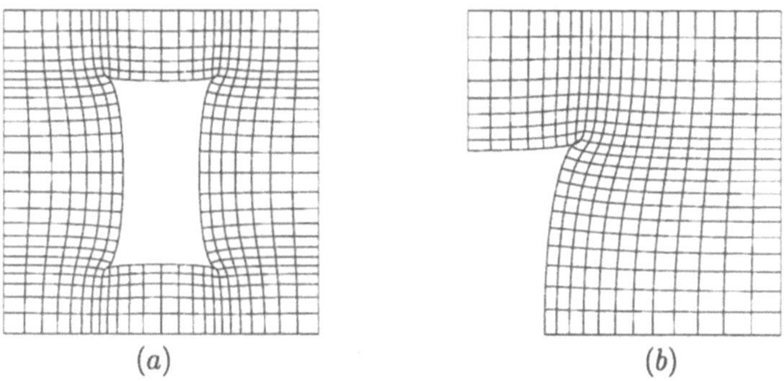

Figure 3.13 *(a): Deformation of the full cell. (b): Deformation of a quarter of the base cell.*

case b:

As with case (a), because of the symmetry of the geometry and loading, we adopt the same boundary conditions. See Figure 3.12. Deformation of the unconstrained cell is depicted in Figure 3.11(c). In Figure 3.14 the deformation of the full cell and a quarter cell are illustrated.

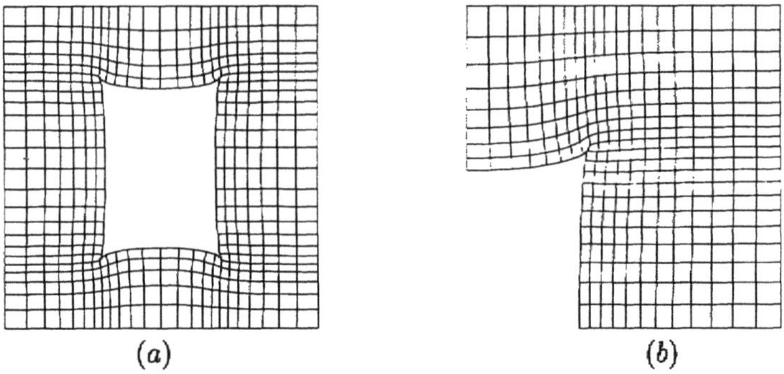

Figure 3.14 (a): Deformation of the full cell. (b): Deformation of a quarter of the base cell.

case c:

Recalling (3.116), the loading in this case is a unit initial shear strain ($\epsilon_{11}^0 = 0$, $\epsilon_{22}^0 = 0$ and $2\epsilon_{12}^0 = 1$). The unconstrained deformation of the cell is shown in Figure 3.11(d). In this case the anti-symmetry condition with respect to the axes of symmetry (s_1 and s_2 in Figure 3.7) exists. This can be verified by noticing that the loading is uniform all over the domain of the cell and for every element in one quadrant of the cell there are three similar ones in the other quadrants which exhibit the anti-symmetry condition.

From the anti-symmetry condition, the displacement of the opposite edges of the base cell may be expressed as

$$u(y_1^0, y_2) = u(y_1^0 + Y_1, y_2), \tag{3.131}$$

$$v(y_1^0, y_2) = -v(y_1^0 + Y_1, y_2), \tag{3.132}$$

$$u(y_1, y_2^0) = -u(y_1, y_2^0 + Y_2), \tag{3.133}$$

$$v(y_1, y_2^0) = v(y_1, y_2^0 + Y_2). \tag{3.134}$$

Comparing (3.131)-(3.134) with (3.121) and (3.122) results in

$$v(y_1^0, y_2) = v(y_1^0 + Y_1, y_2) = 0 \tag{3.135}$$

$$u(y_1, y_2^0) = u(y_1, y_2^0 + Y_2) = 0. \tag{3.136}$$

From (3.135) we observe that for the points located on the left and right hand side edges of the base cell the vertical displacement is zero. According to (3.136) the nodes located on the bottom and top edges have zero horizontal displacement. These boundary conditions are illustrated in Figure 3.15(a).

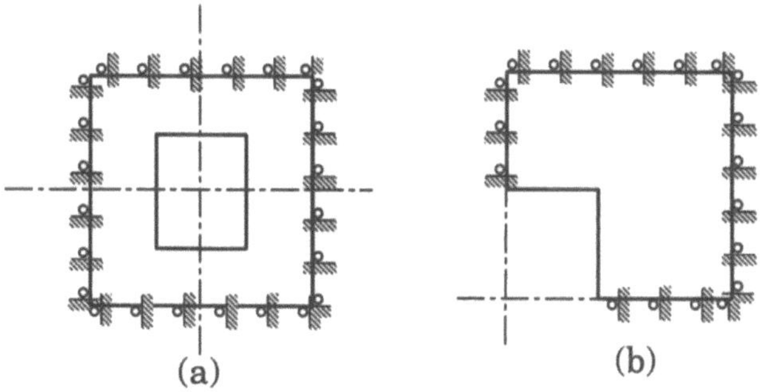

Figure 3.15 *Boundary conditions for case* c

Using the anti-symmetry conditions it is only necessary to consider a quarter of the cell with appropriate boundary conditions as depicted in Figure 3.15(b). As with cases (a) and (b), by putting $y_1 = y_1^0$ and $y_2 = y_2^0$ in (3.135), (3.136), (3.121) and (3.122) and comparing them, we again end up with (3.129) and (3.130) which means that the outer corner points of the cell are fixed and the rigid body mode, as depicted in Figure 3.9(b), does not appear. Deformation of the base cell and a quarter of it for this loading case are illustrated in Figure 3.16.

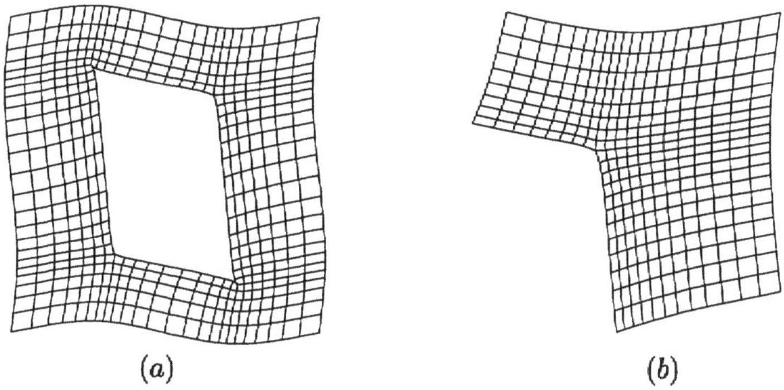

Figure 3.16 *(a): Deformation of the full cell. (b): Deformation of a quarter of the base cell.*

We note that because of similarities of the boundary conditions in cases (a), (b) and (c), one can define the boundary conditions for only case (a) and for case (c) instead of defining a new problem a few lines can be added to the FE code to change it automatically, such as the following pseudocode in FORTRAN

```
IF CASE IS (c) THEN
    IF ( IFPRE(IVFIX(1)) .NE. IFPRE(IVFIX(2)) ) THEN
        ITEMP = IFPRE(IVFIX(1))
        IFPRE(IVFIX(1)) = IFPRE(IVFIX(2))
```

```
        IFPRE(IVFIX(2)) = ITEMP
    END IF
END IF
```

where IFPRE is an indicator to distinguish the restrained degrees of freedom, IVFIX is the number of the fixed node and ITEMP is a dummy variable.

3.4.3 Examples

For the numerical solution of the homogenization equation by the finite element method the included program HOMOG is used.

Example 1: The first example, consists of a unit square base cell containing a 0.4×0.6 void (see Figure 3.17) and with solid phase material properties $D_{11} = D_{22} = 30$ and $D_{12} = D_{66} = 10$ (all units are assumed to be consistent). This problem has been solved by Bendsøe and Kikuchi [34].

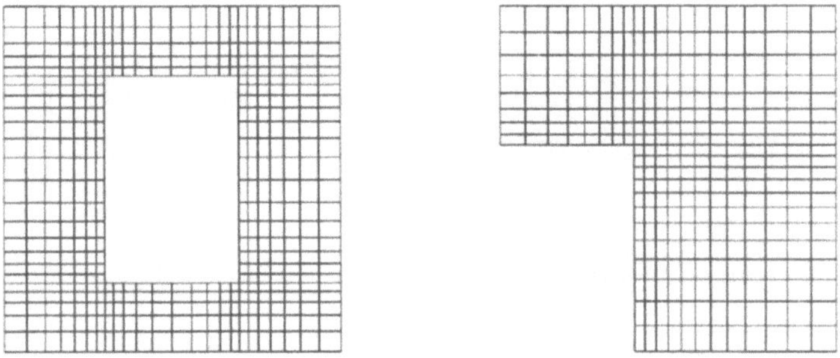

(a) (b)

Figure 3.17 (a): FE mesh for full cell. (b): FE mesh for a quarter of the base cell.

This example was solved using two different meshes. First (case a) the mesh of Figure 3.17(a) with 436 elements is used. In case (b), by making use of symmetry only a quarter of the cell was meshed with 305 elements as shown in Figure 3.17(b). When the problem was executed using one quarter of the mesh of Figure 3.17(a), exactly the same results as case (a) were obtained. In all cases, 8-node elements were adopted. In Table 3.1 the results are compared

Mesh	D^H_{11}	D^H_{12}	D^H_{22}	D^H_{66}	Remarks
20 × 20 4-node	13.015	3.241	17.552	2.785	Reference [34]
1st adapt.	12.910	3.178	17.473	2.714	Reference [34]
2nd adapt.	12.865	3.146	17.437	2.683	Reference [34]
3rd adapt.	12.844	3.131	17.421	2.668	Reference [34]
436 8-node	12.839	3.139	17.422	2.648	HOMOG case (*a*)
305 8-node	12.820	3.124	17.407	2.634	HOMOG case (*b*)

Table 3.1 *Comparison of results*

with the solution of Bendsøe and Kikuchi. They used an initial 20 × 20 FE mesh and then applied the *h*-adaptation method three times.

Example 2: The problem of a cellular material with a perfect honeycomb base cell is now considered. The base cell is illustrated in Figure 3.18. To ensure that the area of the base cell is unity, the dimension $a = 3^{-3/4}$ is chosen. In this example the ratio of the wall thickness to cell size is $t/a = \sqrt{3}/6$. The material is assumed to be isotropic with Young's modulus $E = 0.91$ and the Poisson's ratio $\nu = 0.3$. All units are assumed to be consistent.

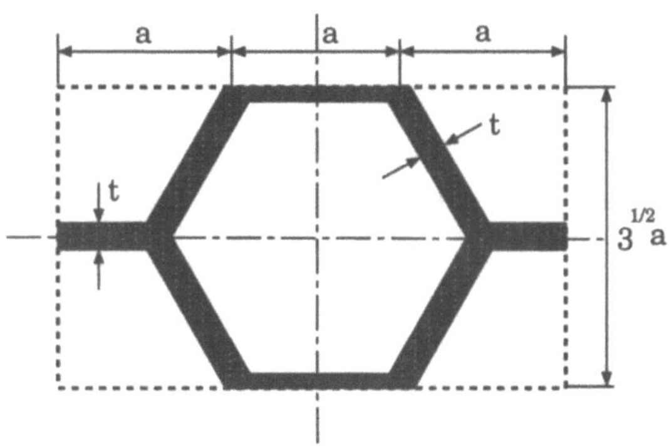

Figure 3.18 *Base cell of a composite with a honeycomb microstructure*

HOMOG Sigmund [35] has solved this problem considering a very weak material instead of a void, so that ($E_{\text{void}}/E_{\text{solid}} = 10^{-4}$). His results using 9072, 4-node elements are shown in the first row of Table 3.2. To compare his results, he also executed this example with PREMAT2D developed by Guedes [36] with 9520, 4-node elements.

In the last row of Table 3.2 the results obtained from running HOMOG are

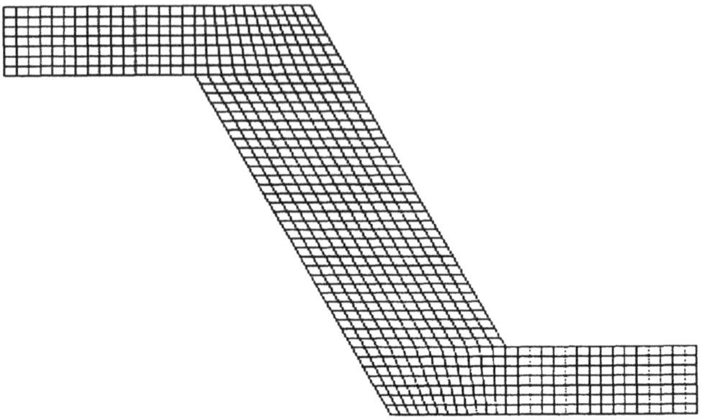

Figure 3.19 FE mesh for a quarter of the honey comb cell

Mesh	D^H_{11}	D^H_{12}	D^H_{22}	D^H_{66}	Remarks
9072 4-node	0.0958	0.0713	0.0950	0.0125	Pixel-like discretization
9520 4-node	0.0968	0.0720	0.0968	0.0124	PREMAT2D
752 9-node	0.0966	0.0720	0.0966	0.0123	HOMOG

Table 3.2 Comparison of results

given. The mesh of Figure 3.19 for only one quarter of the honeycomb cell with 752 9-node bi-quadratic elements is here adopted.

3.4.4 Homogenized constitutive matrix for square microcells with rectangular voids

In order to carry out topology optimization using the model involving material with square microcells with rectangular holes, our aim in numerically solving the homogenization equations is to determine how the constitutive matrix $\mathbf{D}^H$ components vary in terms of the dimensions of the rectangular hole, i.e. to construct $\mathbf{D}^H(a, b)$.

For this purpose a grid of 11×11 equally spaced sampling points for 121 values of a and b was constructed and for each pair of values of a and b a finite element mesh was generated. The material is assumed to be isotropic with $E/(1-\nu^2)$ scaled to equal unity and Poisson's ratio $\nu = 0.3$. The HOMOG program was executed for the generated finite element meshes. In all cases the 8-node serendipity elements were adopted with 3×3 Gauss-Legendre integration. In Table B.1 (Appendix B) the obtained results are given. Variations of the elements of the elasticity matrix with respect to the dimensions of the holes

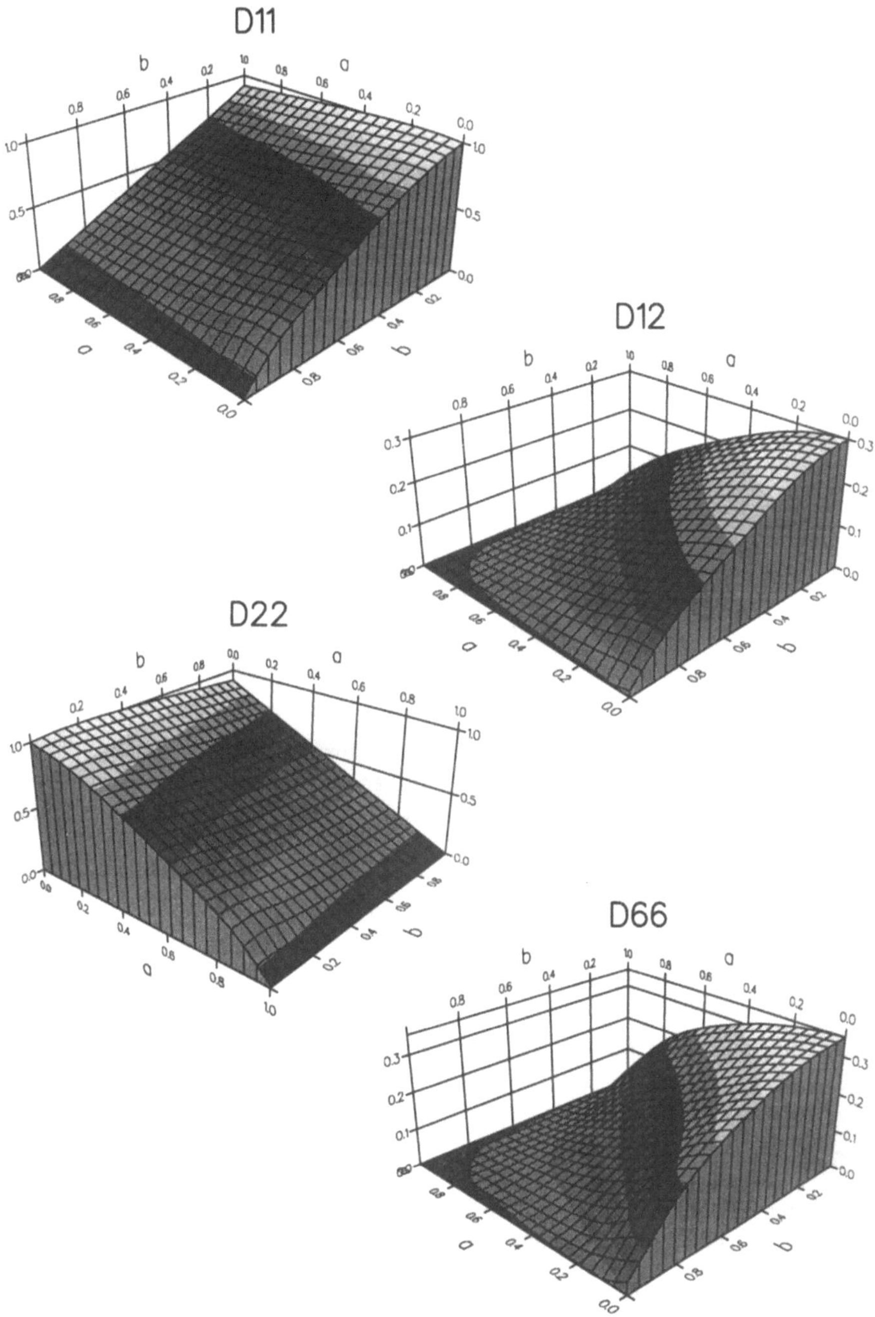

Figure 3.20 *Homogenized elasticity constants with respect to a and b.*

are illustrated in Figure 3.20.

3.4.5 Least squares smoothing

To find the elements of the homogenized elasticity matrix in a polynomial form from the discrete values obtained at the sampling points, different interpolation techniques, such as splines, least squares approximation polynomials or shape functions can be used. Legendre polynomials were suggested by Bendsøe and Kikuchi [11, 34, 37]. Here we employ a conventional least squares data fitting method.

If $D_p^n(a,b)$ is considered as the desirable approximating complete polynomial of degree n, it can be written as

$$D_p^n(a,b) = \sum_{i=0}^{n}\sum_{j=0}^{n} c_k a^i b^j, \qquad k = 1,2,\cdots,\frac{n(n+1)}{2} = N \tag{3.137}$$

where N is the number of terms of the approximating polynomial. The coefficients c_k can be found by minimizing

$$e(c_k) = \sum_{i=1}^{n_i}\sum_{j=1}^{n_j}\left(\hat{D}(a_i,b_j) - D_p^n(a,b)\right)^2 \tag{3.138}$$

where $\hat{D}(a_i,b_j)$ are the elements of the homogenized constitutive matrix at the sampling points and n_i and n_j are the number of points along the a and b axes of the sampling grid (e.g. in Table B.1 $n_i = n_j = 11$). Now from stationary conditions for (3.138) it follows that

$$\frac{\partial e(c_k)}{\partial c_k} = 0\,, \qquad k = 1,2,\cdots,N \tag{3.139}$$

which provides a system of N linear equations from which the coefficients c_k can be calculated.

The results obtained for the complete bi-cubic and bi-quartic approximating polynomials are given in Appendix C.

References

[1] Cheng K.T. and Olhoff N., An investigation concerning optimal design of solid elastic plates, *Int. J. Solids Struct.*, **16**, 305–323 (1982)

[2] Cheng K.T.. and Olhoff N., Regularized formulation for optimal design of axisymetric plates, *Int. J. Solids Struct.*, **18**, 153–170 (1982)

[3] Lurie K.A. and Chirkaev A.V., G-Closure of some particular sets of admissible material characteristics for the problem of bending of thin plates, *J. Optim. Theory Appl.*, **42**, 305–315 (1984)

[4] Rozvany G.I.N., Olhoff N., Bendsøe M.P., Ong T.G., Sandler R. and Szeto W.T., Least-weight design of perforated elastic plates, Parts I and II, *Int. J. Solids Struct.*, **23**, 521–536, 537–550 (1987)

[5] Ong T.G., Rozvany G.I.N. and Szeto W.T., Least-weight design of perforated elastic plates for given compliance: non-zero Poisson's ratio, *Comp. Meth. Appl. Mech. Eng.*, **66**, 301–322 (1988)

[6] Kohn R.V. and Strang G., Optimal design for torsional rigidity, in *Hybrid and mixed finite element methods*, edited by Atluri S.N., Gallagher R.H. and Zienkiewicz O.C., pp. 281–288. Wiley & Sons (1983)

[7] Rozvany G.I.N., Zhou M. and Birker T., Generalized shape optimization without homogenization, *Structural Optimization*, **4**, 250–252 (1992)

[8] Kohn R.V. and Strang G., Optimal design and relaxation of variational problems, *Comm. Pure Appl. Math.*, **39**, 113–137 (Part I), 139–182 (Part II), 353–377 (Part III) (1986)

[9] Bendsøe M.P., Optimal shape design as a material distribution problem, *Structural Optimization*, **1**, 193–202 (1989)

[10] Bonnetier E. and Vogelius M., Relaxation of a compliance fuctional for a plate optimization problem, Tech. rep., University of Maryland (1986)

[11] Bendsøe M.P., Díaz A.R. and Kikuchi N., Topology and generalized layout optimization of elastic structures, in *Topology design of structures*, edited by Bendsøe M.P. and Mota Soares C. A., pp. 159–205. Kluwer Academic Publishers (1993)

[12] Avellaneda M., Optimal bounds and microgeometries for elastic two-phase composites, *SIAM J. Appl. Math.*, **47**, 1216–1228 (1987)

[13] Avellaneda M. and Milton G.W., Bounds on the effective elasticity tensor composites based on two-point correlation, in *Composite Material Technology*, edited by Hui D. and Kozik T.J., pp. 89–93. ASME (1989)

[14] Avellaneda M., Bounds on the effective elastic constants of two-plane composite materials, in *Proc. Sem. College de France*, edited by Brezis X. and Lions J. L. Pitman, London (1989)

[15] Allaire G. and Kohn R.V., Optimal bounds on the effective behavior of a mixture of two well ordered elastic materials, *Quar. Appl. Math.* (1992)

[16] Olhoff N., Krog L. and Thomsen J., Bi- material topology optimization, in *Structural Optimization 93*, edited by Herskovites J., vol. 1 (1993)

[17] Thomsen J., Topology optimization of structures composed of one or two materials, *Structural Optimization*, **5**, 108–115 (1992)

[18] Maute K. and Ramm E., Adaptive techniques in topology optimization, in *Proceedings of Symposium on Multidisciplinary Analysis and Optimization; Panama City, Florida, USA* (1994)

[19] Rozvany G.N., Bendsøe M.P. and Kirsch U., Layout optimization of structures, *Appl. Mech. Rev.*, **48**(2), 41–119 (1995)

[20] Zhou M. and Rozvany G.I.N., The COC algorithm, part II: Topological, geometrical and generalized shape optimization, *Comp. Meth. Appl. Mech. Eng.*, **89**, 309–335 (1991)

[21] Maute K. and Ramm E., The genesis of continuum structures, in *Proceedings of 3rd International Symposium of the SFB 230; Stuttgart* (1994)

[22] Maute K. and Ramm E., Topology optimization of plate and shell structures, in *Proceedings of IASS-ASCE International Symposium; Atlanta, Georgia, USA* (1994)

[23] Ramm E., Bletzinger K.U., Reitinger R. and Maute K., The challenge of structural optimization, in *Advances in structural optimization*, edited by Topping B.H.V. and Papadrakakis M., pp. 27–52. CIVIL-COMP Ltd (1994)

[24] Rozvany G.I.N (ed.), *Shape and layout optimization of structural systems and optimality criteria methods.* Springer-Verlag, CISM, Udine (1992)

[25] Hinton E. and Owen D.R.J., *An introduction to finite element computations.* Pineridge Press Limited, Swansea, U.K. (1979)

[26] Zienkiewicz O.C. and Taylor R.L., *The finite element method*, vol. 1. McGraw-Hill, London, 4th edn. (1989)

[27] Hughes T.J.R., *The finite element method.* Prentice-Hall International, Inc., Englewood Cliffs, New Jersey (1987)

[28] Hinton E. and Owen D.R.J., *Finite Element Programming.* Academic Press, London (1977)

[29] Zienkiewicz O.C. and Scott F.C., On the principle of repeatability and its application in analysis of turbine and pump impellers, *Int. J. Num. Meth. Eng.*, **4**, 445–452 (1972)

[30] Cook R.D., Malkus D.S. and Plesha M.E., *Concepts and application of finite element analysis.* John Wiley & Sons, New York, 3rd edn. (1989)

[31] Sigmund O., *Materials with prescribed constitutive parameters: an inverse homogenization problem*, Tech. Rep. 470, The Tech. Univ. of Denmark (1993)

[32] Hassani B., Finite element solution of the homogenization equation for symmetric cells of periodicity, Tech. Rep. CR/887/95, Dept. of Civil Eng. (1995)

[33] Hassani B., A direct method to derive the boundary conditions of the homogenization equation for symmetric cells, *Comm. Numer. Meth. Eng.*, **12**, 185–196 (1996)

[34] Bendsøe M.P. and Kikuchi N., Generating optimal topologies in structural design using homogenization method, *Comp. Meth. Appl. Mech. Eng.*, **71**, 197–224 (1988)

[35] Sigmund O., *Design of material structures using topology optimization*, Ph.D. thesis, Tech. Univ. of Denmark (1994)

[36] Guedes J.M. and Kikuchi N., Pre and post processing for materials based on the homogenization method with adaptive finite element methods, *Comp. Meth. Appl. Mech. Eng.*, **83**, 143–198 (1990)

[37] Kikuchi N. and Suzuki K., Structural optimization of a linearly elastic structure using the homogenization method, in *Shape and Layout Optimization in Structural Design*, edited by Rozvany G. (1990)

Part II

Topology Optimization

4

STRUCTURAL TOPOLOGY OPTIMIZATION USING OPTIMALITY CRITERIA METHODS

In this chapter, the basic concepts related to the optimality criteria methods are introduced. Then the mathematical model for the structural topology optimization problem is constructed. The optimality conditions are explained and an optimization procedure based on optimality criteria methods is presented. Resizing schemes for updating design variables are explained. The issue of the optimal orientation for the homogenized orthotropic material is discussed. The algorithm of the computer program PLATO is explained and a few illustrative examples are presented.

4.1 Introduction

In Chapter 1 a mathematical description of the general optimization problem was presented. In Chapter 3 by introducing different material models, we noticed how the complex nature of the structural topology (or layout) optimization problem can be changed to an easier sizing problem. The design variables of the optimization problem are the geometrical parameters defining the microstructures of the assumed composite material in each finite element.

To solve the topology optimization problem any non-linear mathematical programming method may be used. These methods are usually robust and can be applied to any sort of optimization problem. However, when applied to structural optimization problems several calculations of the objective function, constraint functions and their derivatives (i.e. several structural and sensitivity analyses) are usually needed [1, 2]. Because of this, the required computer time for mathematical programming methods is largely dependent on the number of design variables, which means that when the number of design variables increases, these methods become very costly and time consuming [3].

In structural topology optimization based on the homogenization concept, the number of design variables is proportional to the number of elements of the discretized domain. Consequently, the use of mathematical programming methods for realistic topology optimization problems is somewhat impractical, as they usually require hundreds of finite elements, and therefore hundreds of design variables. To overcome this deficiency 'optimality criteria' methods have been considered.

Optimality criteria methods were developed in the late 1960's as an alternative approach to mathematical programming methods for solving structural optimization problems. The fundamental ideas behind optimality criteria methods were originally introduced by Michell in 1904 [4]. An excellent review of the history of optimality criteria methods is given in Chapter 9 of a book by Rozvany [5].

The optimality criteria methods are indirect methods of optimization and unlike mathematical programming methods which directly optimize the objective function, optimality criteria methods attempt to satisfy a set of criteria related to the behaviour of the structure. These criteria are derived either intuitively or rigorously. 'Fully stressed design' and 'simultaneous failure mode design' are examples of intuitive optimality criteria methods. Rigorously derived optimality criteria methods, often based on Kuhn-Tucker optimality conditions, are the subject of this chapter.

Optimality criteria methods for structural optimization have been used by two groups of researchers: the so-called analytical and numerical schools [6, 7]. The newly developed, discretized, continuum-type optimality criteria technique (DCOC) [8, 7] takes advantage of the approaches adopted in both schools and is more useful than continuum-based optimality criteria (COC) methods [7, 9].

The COC methods are usually derived by variational methods. Rozvany and his coworkers [5, 6, 10] reinterpret the derived optimality criteria as equilibrium, compatibility and generalized strain-stress relations for a fictitious system called the *adjoint structure*. This fictitious structure is used as a convenient mechanical analogy for interpreting certain quantities in optimality criteria. This interpretation has two advantages. First, engineers can visualize it more easily than an abstracted mathematical problem and second, the analysis of the adjoint structure, as well as the actual structure, can be carried out with the finite element method [10]. In iterative COC methods applied to discretized problems, each iteration consists of two basic steps: first, the analysis of the real and adjoint systems and second, the update of the design variables [10].

The rapidly increasing computational capacity of modern computers and the necessity for discretization in most practical problems have provided the main motives for DCOC methods. In the DCOC method developed by Rozvany and his coworkers, the general optimality criteria for discretized structural systems are derived on the basis of the flexibility formulation of matrix analysis and the Kuhn-Tucker optimality conditions. An iterative algorithm for coupling the DCOC method with finite element analysis is then developed [9].

Optimality criteria methods are especially efficient for problems with large numbers of design variables and few constraints. These criteria are, in fact, necessary (and sometimes sufficient) conditions for minimality of the objective function.

Although closed form analytical solutions can be achieved for very limited and idealized systems, they are very important in checking the validity and applicability of numerical methods as well as discovering the fundamental features of optimal solutions. Hence, in the following, two simple illustrative examples of analytical optimality criteria will be presented.

4.2 Kuhn-Tucker optimality conditions

The optimization problem in its general form can be expressed as follows

$$\begin{aligned} &\text{Minimize} && f(\mathbf{x}) && \\ &\text{such that} && h_j(\mathbf{x}) = 0, && j = 1, 2, \cdots, n_h \\ & && g_k(\mathbf{x}) \leq 0, && k = 1, 2, \cdots, n_g \end{aligned} \tag{4.1}$$

where $\mathbf{x} = [x_1, x_2, \cdots, x_n]T$ are design variables and n_h and n_g are respectively the number of equality and inequality constraints.

The inequality constraints can be transformed into equality constraints by adding *slack variables*. That is, the inequality constraints can be written as

$$g_k(\mathbf{x}) + s_k^2 = 0, \qquad k = 1, 2, \cdots, n_g. \tag{4.2}$$

The necessary conditions for a minimum of a function subjected to some equality constraints can be obtained by using the classical technique of Lagrange multipliers. Typically, the Lagrangian function is written as

$$\mathcal{L}(\mathbf{x},\mathbf{s},\boldsymbol{\lambda},\boldsymbol{\nu}) = f(\mathbf{x}) + \sum_{j=1}^{n_h} \lambda_j h_j(\mathbf{x}) + \sum_{k=1}^{n_g} \nu_k \left(g_k(\mathbf{x}) + s_k^2 \right) . \qquad (4.3)$$

where λ_j and ν_k are Lagrangian multipliers. Stationary conditions for $\mathcal{L}$ are equivalent to the original problem (4.1). That is

$$\frac{\partial \mathcal{L}}{\partial x_i} = \frac{\partial f}{\partial x_i} + \sum_{j=1}^{n_h} \lambda_j \frac{\partial h_j}{\partial x_i} + \sum_{k=1}^{n_g} \nu_k \frac{\partial g_k}{\partial x_i} = 0 \qquad i = 1,\cdots,n \qquad (4.4)$$

$$\frac{\partial \mathcal{L}}{\partial \lambda_j} = h_j = 0 \qquad j = 1,\cdots,n_h \qquad (4.5)$$

$$\frac{\partial \mathcal{L}}{\partial \nu_k} = g_k + s_k^2 = 0 \qquad k = 1,\cdots,n_g \qquad (4.6)$$

$$\frac{\partial \mathcal{L}}{\partial s_k} = 2\,\nu_k\, s_k = 0 \qquad k = 1,\cdots,n_g . \qquad (4.7)$$

Equations (4.5) are the same as the equality constraints in (4.1). Furthermore, (4.6) ensure that $g_k \le 0$. From (4.7), known as *switching conditions*, it is concluded that either ν_k or s_k is zero (or both). The condition $s_k = 0$ states that the constraint g_k is active. However, $s_k \ne 0$ implies that $\nu_k = 0$ and g_k is inactive which means that it can be removed without affecting the solution. If both $s_k = 0$ and $\nu_k = 0$ (sometimes called an *abnormal case* [11]), then this means that the surface $g_k = 0$ passes through the minimum obtained by the rest of problem and the other constraints.

From (4.6) and (4.7) we conclude that

$$g_k \le 0 \qquad \text{and} \qquad \nu_k\, g_k = 0 . \qquad (4.8)$$

We observe that by replacing (4.6) and (4.7) with (4.8) we do not, in fact, lose any information.

Equation (4.4) may be written as

$$\nabla f + \sum_{j=1}^{n_h} \lambda_j \nabla h_j + \sum_{k=1}^{n_{ga}} \nu_k \nabla g_k = 0 , \qquad (4.9)$$

where n_{ga} ($n_{ga} \le n_g$) is the number of active constraints. From (4.9) it is observed that ∇f lies in the subspace spanned by ∇h_j and those of ∇g_k which correspond to the active constraints. In other words ∇f must be expressible as a linear combination of the normals to the surfaces $h_j = 0$ and those of g_k which are active.

So far, by using the theorem of Lagrange multipliers, which ensures the existence of the set of real numbers λ_j and ν_k, we have found a system of nonlinear equations (4.4)-(4.7) to be solved. These are the necessary conditions for a point to be stationary.

Now we are ready to explain the Kuhn-Tucker conditions. In 1951 the theorem of Kuhn and Tucker [12] provided one additional condition to the above conditions which are derived from the slack variable approach and the use of the Lagrange multipliers. However, these extra conditions, called Kuhn-Tucker conditions, are valid only for problems constrained by inequalities.

Luckily, converting the general problem (4.1) to an equivalent problem without equality constraints is straightforward, since any equality relationship can always be converted to inequalities. For example, $h = 0$ can be replaced by the two inequalities $h \leq 0$ and $-h \leq 0$. Another approach, which is more practical, is to use the Lagrange multiplier technique at the first stage to replace the objective function together with the equality constraints with a new objective function in the Lagrangian form (e.g. $F(\mathbf{x}, \boldsymbol{\lambda})$). Therefore without any loss of generality and for the sake of simplicity the optimization problem may be stated as

$$\begin{aligned} &\text{Minimize} && F(\mathbf{x}) \\ &\text{subject to} && g_k(\mathbf{x}) \leq 0, \qquad k = 1, 2, \cdots, n_g. \end{aligned} \tag{4.10}$$

According to the Kuhn-Tucker theorem, if $\mathbf{x}$ is a local minimum of $F(\mathbf{x})$ which satisfies the constraints, then there exist multipliers ν_k such that

$$\begin{aligned} &\nabla F(\mathbf{x}) + \sum_{k=1}^{n_g} \nu_k \nabla g_k(\mathbf{x}) = \mathbf{0}\ , \\ &\nu_k\, g_k(\mathbf{x}) = 0\ , && k = 1, \cdots, n_g \\ &\nu_k \geq 0\ , && k = 1, \cdots, n_g. \end{aligned} \tag{4.11}$$

The inequality and the two equalities in (4.11) are called *Kuhn-Tucker conditions.* The difference between the Kuhn-Tucker conditions and what we get from the Lagrange multipliers method, is that ν_k in (4.11) is sign definite (non-negative), while the Lagrange multipliers theorem only ensures the existence of ν_k and does not give any extra information about its sign.

As we have noticed earlier, ν_k for inactive constraints ($g_k(\mathbf{x}) \neq 0$) is zero. Consequently, (4.11) may be summarized for that case as

$$\begin{aligned} &\nabla F(\mathbf{x}) + \sum_{k=1}^{n_{ga}} \nu_k \nabla g_k(\mathbf{x}) = \mathbf{0}\ , \\ &\nu_k \geq 0\ . \end{aligned} \tag{4.12}$$

The Kuhn-Tucker conditions (4.12) for a regular point[†] state that $-\nabla F$ has

[†] A point is said to be a regular point of the design space if the gradients of all active constraints at that point are linearly independent.

to fall within the convex cone comprising the set of all non-negative linear combinations of the gradients of active constraints. This, in fact, ensures that it is not possible to find a move vector in the feasible region with a component on $-\boldsymbol{\nabla} F$ which is the steepest descent direction for the objective function $F(\mathbf{x})$. A two dimensional geometric interpretation is illustrated in Figure 4.1 .

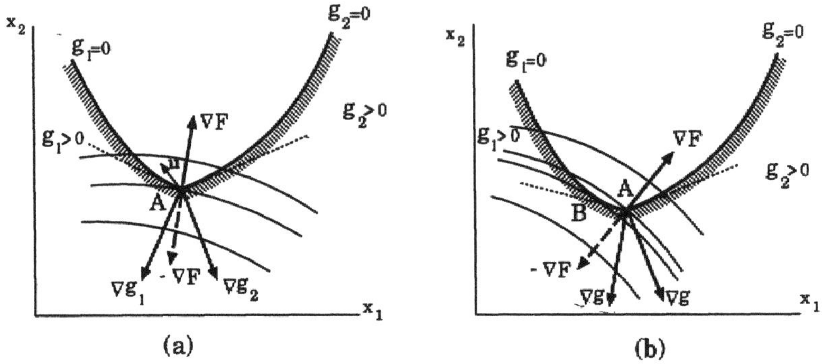

Figure 4.1 (a) *Kuhn-Tucker conditions are satisfied and A is a minimum.* (b) *As $-\boldsymbol{\nabla} F$ is not within the cone of gradients of the active constraints, Kuhn-Tucker conditions are not satisfied and A is not minimum.*

Assume that we are going to test point A, shown in Figure 4.1, to see whether it is a minimum or not. If A is not the local minimum, then we have to be able to find a move vector $\mathbf{u}$ which is inside the feasible region so that by moving along it the objective function decreases. From $\mathbf{u}$ being feasible it follows that

$$\mathbf{u}^T \boldsymbol{\nabla} g_k \leq 0 \tag{4.13}$$

and $\mathbf{u}$ being usable requires that

$$\mathbf{u}^T(-\boldsymbol{\nabla} F) > 0 . \tag{4.14}$$

By substituting for $-\boldsymbol{\nabla} F$ from the first equation of (4.11), (4.14) can be written as

$$\mathbf{u}^T \left(\sum_{k=1}^{n_{ga}} \nu_k \boldsymbol{\nabla} g_k \right) > 0 \tag{4.15}$$

which is the same as

$$\sum_{k=1}^{n_{ga}} \nu_k \left(\mathbf{u}^T \boldsymbol{\nabla} g_k \right) > 0 . \tag{4.16}$$

From (4.13) and (4.16) it is concluded that the condition for A to be a local minimum is that $\nu_k \geq 0$ (which makes (4.16) impossible).

Using the Kuhn-Tucker conditions, we can establish a test to be applied to a candidate point to see whether that point is a minimum or not, instead of solving the set of nonlinear equations. For convex minimization problems† , the necessary Kuhn-Tucker conditions, are also sufficient. Furthermore, any local minimum is a global one.

Here, as the final point, the necessary optimality conditions for the optimization problem in its general form of (4.1) is summarized, that is:

$$\begin{aligned}
&\frac{\partial f}{\partial x_i} + \sum_{j=1}^{n_h} \lambda_j \frac{\partial h_j}{\partial x_i} + \sum_{k=1}^{n_g} \nu_k \frac{\partial g_k}{\partial x_i} = 0 && i = 1, \cdots, n \\
&h_j = 0 && j = 1, \cdots, n_h \\
&g_k \leq 0 && j = 1, \cdots, n_g \\
&g_k \, \nu_k = 0 && k = 1, \cdots, n_g \\
&\nu_k \geq 0 && k = 1, \cdots, n_g .
\end{aligned} \tag{4.17}$$

It is important to note that by changing the optimization problem from finding a minimum to a maximum, or by changing the sign of constraint terms in the Lagrangian (4.3), or by changing the direction of the inequality constraints, each time, the sign of multipliers ν_k in the Kuhn-Tucker conditions (4.11) will change [13].

4.3 Analytical optimality criteria

As mentioned in the introduction of this chapter, analytical optimality criteria methods, apart from having an important role in the development of OC methods, are used for the verification of numerical algorithms. In this section two illustrative examples are considered. Our objective in the first example is to give a concise introduction to the variational method and terminology frequently used by the analytical school [5]. The object of the second example is first, to explain the main aspects of the procedure of the analytical method and second, we will use the results for verification of the algorithm developed for structural topology optimization. It is hoped that this will help engineers engaged with numerical techniques to have better communication with the analytical school.

4.3.1 An illustrative example of variational analysis

As a simple example of the derivation of the analytical optimality criteria for a continuum structure, the Euler-Bernoulli beam problem, with a single criterion

† If the objective function and feasible domain of a problem are convex, it is called a *convex programming* problem. Convexity of the feasible domain occurs if all the inequality constraints are convex and the equality constraints are linear.

is considered here [14]. The height of the cross section $z(x)$ and the deflection of the beam $y(x)$ are assumed to be, respectively, the design variable and response (state) functions. See Figure 4.2. Note that the admissible functions y should satisfy the kinematic boundary conditions. The objective function is given in the form of a global measure as $\int_0^L \rho(x)y(x)dx$, where $\rho(x)$ is a specified weighting on the beam displacement $y(x)$. The lower bound for z and the upper bound for the volume of the beam are respectively denoted by z^o and V. The state (equilibrium) equation of the beam is

$$\frac{d^2}{dx^2}\left(EI(x)\frac{d^2y(x)}{dx^2}\right) - p(x) = 0 \qquad \textit{(and boundary conditions)}. \tag{4.18}$$

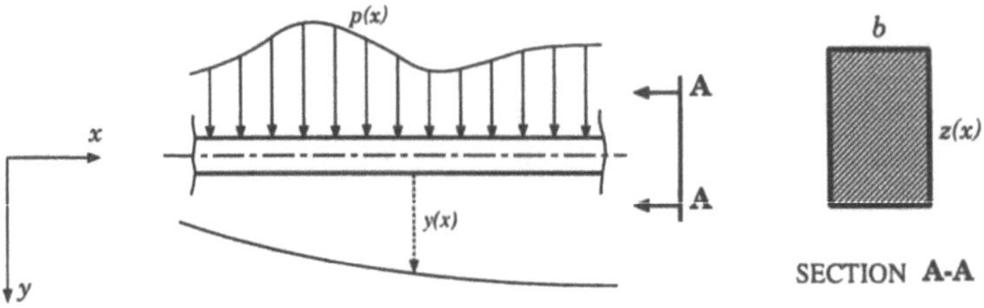

Figure 4.2 Euler-Bernoulli beam

Consequently, the optimal design problem may be constructed as

$$\begin{aligned}
&\text{Minimize} && \int_0^L \rho(x)y(x)dx \\
&\text{subject to} && \left(S(x,z(x))y''(x)\right)'' - p(x) = 0 \\
& && z^o - z(x) \le 0 \\
& && \int_0^L bz(x)dx - V \le 0
\end{aligned} \tag{4.19}$$

where L is the length of the beam, S is the bending rigidity (EI) and the superscript $(')$ implies differentiation with respect to x. By using Kuhn-Tucker conditions as explained in the last section, the Lagrangian of (4.19) may be written as

$$\mathcal{L} = \int_0^L \rho(x)y(x)dx + \int_0^L \bar{y}(x)\left[\left(S(x,z(x))y''\right)'' - p(x)\right]dx +$$
$$\int_0^L \gamma(x)\left[z^o - z(x)\right]dx + \Gamma\left[\int_0^L bz(x)dx - V\right], \tag{4.20}$$

where $\bar{y}(x), \gamma(x)$ and Γ are Lagrange multipliers. From stationarity of $\mathcal{L}$ with respect to y and by cancelling the terms with opposite signs in the related

Euler-Lagrange equation (or equivalently by integrating by parts from the second term of (4.20) four times), it follows that

$$\Big(S\big(x, z(x)\big)\bar{y}''(x)\Big)'' - \rho(x) = 0\,. \tag{4.21}$$

This equation is very similar to the state equation (4.18) and the only difference is the loading. Because of this similarity it is called an *adjoint system*. Similarly, stationarity of $\mathcal{L}$ with respect to $z(x)$ results in

$$\frac{dS(x,z)}{dz}\bar{y}''y'' + \gamma(x) - \Gamma b = 0. \tag{4.22}$$

This equation is referred to as the *optimality criterion*. The term $dS/dz\bar{y}''y''$ is, in fact, the *gradient* of the objective function with respect to the design variable $z(x)$†. Using the Kuhn-Tucker conditions the other optimality conditions are obtained; that is

$$z^o - z(x) \le 0\,; \quad \gamma(x)\big(z^o - z(x)\big) = 0\,; \quad \gamma(x) \ge 0 \tag{4.23}$$

and

$$\int_0^L bz(x)dx - V \le 0\,; \quad \Gamma\left[\int_0^L bz(x)dx - V\right] = 0\,; \quad \Gamma \ge 0\,. \tag{4.24}$$

$\frac{dS}{dz}\bar{y}''y''$ is also the *specific mutual energy* between the original and *adjoint* states. We note that over intervals where $z(x) > z^o$, according to (4.23) $\gamma(x) = 0$. Thus, the optimality condition requires that the gradient of the objective function, or the specific mutual energy, has to be of constant value.

If the weighting function $\rho(x)$ in the preceding example equals the primary load (i.e. $\rho(x) = p(x)$), then the solutions of (4.18) and (4.21) are the same, so that $y(x) = \bar{y}(x)$. This implies that the adjoint structure (in terminology used by Rozvany [5, 6, 10]) is the same as the real structure. Also we note that in this case the objective function is a measure of *compliance*.

---

† This can be verified by noticing that the objective function is in fact the external work done by the virtual force $\rho(x)$ on the real displacements $y(x)$, which is equal to the internal work, that is

$$\Phi = \int_0^L \rho y dx = \int_0^L \frac{M\bar{M}}{S}dx = \int_0^L S\bar{y}''y''dx$$

where $M = Sy''$ and $\bar{M} = S\bar{y}''$ are the moments of inertia and differentiating it results in

$$\frac{d\Phi}{dz} = \frac{dS}{dz}\bar{y}''y''.$$

4.3.2 An illustrative example of derivation of optimality criteria

The problem of exact layout optimization of a plane truss with the objective of minimizing the weight of the structure and subject to several load conditions and displacement constraints has been addressed by Rozvany and coworkers [6, 8]. Here the problem of minimum compliance (maximum stiffness) with an upper side volume constraint and a single load case is considered. We will see that the optimality criteria are the same and the result obtained will be used as a benchmark to check the numerical algorithm developed.

Following Rozvany *et al* [6], we first derive the optimality criteria for a finite number of truss elements with lower side constraints for the cross sectional area of elements. The mathematical expression of the problem may be written as: Find the set of cross-sectional areas which

$$\begin{aligned}
&\text{Minimize} && \frac{1}{2}\mathbf{p}^T\mathbf{u} \\
&\text{subject to} && \frac{1}{2}\mathbf{p}^T\mathbf{u} = \frac{1}{2}\sum_{i=1}^{n}\frac{f_i^2\ell_i}{a_iE_i} \\
& && a_i^o - a_i \leq 0\,; \qquad i = 1,\cdots,n \\
& && \sum_{i=1}^{n} a_i\ell_i - V \leq 0\,,
\end{aligned} \tag{4.25}$$

where $\mathbf{p}$ and $\mathbf{u}$ are respectively the external load and the related displacement vectors. n is the number of elements, f_i is the internal force in element i and a_i, ℓ_i and E_i are respectively the cross-sectional area, length and modulus of elasticity of member i of the truss. The symbol V stands for the volume and a_i^o is the lower bound of the cross-sectional area. The optimization problem (4.25) can also be simplified as

$$\begin{aligned}
&\text{Minimize} && \sum_{i=1}^{n}\frac{f_i^2\ell_i}{a_iE_i} \\
&\text{subject to} && a_i^o\ell_i - a_i\ell_i \leq 0\,; \qquad i = 1,\cdots,n \\
& && \sum_{i=1}^{n} a_i\ell_i - V \leq 0\,.
\end{aligned} \tag{4.26}$$

The Lagrange function may be constructed as

$$\mathcal{L} = \sum_{i=1}^{n}\frac{f_i^2\ell_i}{a_iE_i} + \sum_{i=1}^{n}\lambda_i(a_i^o\ell_i - a_i\ell_i) + \Gamma\left(\sum_{i=1}^{n} a_i\ell_i - V\right), \tag{4.27}$$

where λ_i and Γ are the Lagrange multipliers. Employing the Kuhn-Tucker conditions it follows that

$$-\frac{f_i^2\ell_i}{a_i^2E_i} - \lambda_i\ell_i + \Gamma\ell_i = 0\,, \qquad i = 1,2,\cdots,n \tag{4.28}$$

$$\lambda_i \geq 0 \; ; \quad \lambda_i(a_i^o - a_i) = 0, \qquad i = 1, 2, \cdots, n \tag{4.29}$$

$$\Gamma \geq 0 \; ; \quad \Gamma \left(\sum_{i=1}^{n} a_i \ell_i - V \right) = 0 \, . \tag{4.30}$$

It is important to note, however, that in this problem Γ in (4.28) and (4.30) can not be zero and it is a strictly positive number. This can be proven by considering (4.28) where assuming $\Gamma = 0$ results in

$$\frac{f_i^2 \ell_i}{a_i E_i} = -\lambda_i a_i \ell_i . \tag{4.31}$$

According to (4.29) $\lambda_i \geq 0$ and it can be concluded that

$$\frac{1}{2} \frac{f_i^2 \ell_i}{a_i E_i} \leq 0 \, . \tag{4.32}$$

But the left hand side of (4.32) is the strain energy of element i of the truss which can not be negative or zero which implies infinite rigidity. This is clearly impossible. The cross sectional area can be obtained from (4.28) as

$$a_i = \sqrt{\frac{f_i^2}{(\Gamma - \lambda_i) E_i}} \, . \tag{4.33}$$

Therefore, from the switching conditions (4.29) it follows that

$$\begin{cases} \text{if} \quad a_i > a_i^o & \text{then } \lambda = 0 \quad \text{and} \quad a_i = \sqrt{\dfrac{f_i^2}{\Gamma E_i}} \\ \text{if} \quad a_i = a_i^o & \text{then } \lambda \geq 0 \quad \text{and} \quad a_i \leq \sqrt{\dfrac{f_i^2}{\Gamma E_i}} \, . \end{cases} \tag{4.34}$$

Using the definition of strain $\epsilon = f_i / a_i E_i$ and substituting into (4.34) results in

$$\begin{cases} \sqrt{\frac{E_i}{\Gamma}} \, \epsilon_i = 1 & \text{for} \quad a_i > a_i^o \\ \sqrt{\frac{E_i}{\Gamma}} \, \epsilon_i \leq 1 & \text{for} \quad a_i = a_i^o \, . \end{cases} \tag{4.35}$$

If we define a *criterion function* Φ as

$$\Phi = \sqrt{\frac{E_i}{\Gamma}} \, \epsilon_i \tag{4.36a}$$

then (4.35) can be reduced to

$$\begin{cases} \Phi = 1 & \text{for} \quad a_i > a_i^o \\ \Phi \leq 1 & \text{for} \quad a_i = a_i^o \, . \end{cases} \tag{4.36b}$$

Now, we extend the derived criteria to vanishing cross sections by letting $a_i \to 0$. Note that Γ is still greater than zero because $\Gamma = 0$ leads us to a truss with infinite rigidity. In this case (4.36b) can be written as

$$\begin{cases} \Phi = 1 & \text{for} \quad a_i > 0 \\ \Phi \leq 1 & \text{for} \quad a_i = 0 \, . \end{cases} \tag{4.37}$$

Now, we are in the position to extend the result obtained to the 'structural universe' with an infinite number of elements, i.e. to a continuum comprising an infinite number of truss-like elements in all potential directions. In this case, at each point (x, y) of the domain of interest a potential truss member passing through that point, can be represented by the coordinates of the point and the orientation θ of the member. Therefore, the strain in each member (ϵ_i) can well be identified by $\epsilon(x, y, \theta)$. Hence, (4.37) in this case is changed to

$$\begin{cases} \Phi(x,y,\theta) = 1 & \text{for} \quad a(x,y,\theta) > 0 \\ \Phi(x,y,\theta) \leq 1 & \text{for} \quad a(x,y,\theta) = 0 \,. \end{cases} \tag{4.38}$$

So far, we have come to the conclusion that the optimality criteria for the problem of a minimum weight truss with some constraints on the displacements and stresses (which has been addressed by many researchers) and the problem of minimum compliance with a side constraint on the volume (or weight) are identical. The only difference is that in the latter the real and adjoint structure (see Reference [6]) are the same.

Now, following a procedure similar to the one introduced by Rozvany *et al* [6], we argue that $\Phi = \epsilon(x,y,\theta)\sqrt{E_i/\Gamma}$ has a maximum value in the direction of non-vanishing (here this implies optimal) members. Thus, at each point (x, y) there exists an optimal passing member in the direction θ^* only if

$$|\epsilon(x,y,\theta^*)| = \max_{\theta} |\epsilon(x,y,\theta)| \tag{4.39}$$

where θ^* is a specific value of θ.

It is interesting to notice that the 'continuum universe' of the problem can equivalently be thought of as a two (or three) dimensional elasticity problem. Consequently, for a kinematically admissible displacement field, the direction of the maximum principal strain at each point matches the direction of the optimal member. Furthermore, from (4.36a) it is observed that the optimality criteria function is proportional to the strain. Therefore, by normalizing our problem (i.e. adopting $\sqrt{E/\Gamma} = 1$ which does not affect the optimum layout), we can conclude that the condition for optimal members is $|\epsilon_1| = 1$. We note that the direction of the minor principal strain is orthogonal to the direction of the major principal strain and in general $|\epsilon_2| \leq 1$.

Now, let us consider the problem of a single vertical load in Figure 4.3 which is to be optimally transmitted to a vertical support. The kinematic boundary conditions are

$$u(0,y) = v(0,y) = 0 \,, \tag{4.40}$$

where u and v are displacements in the x and y directions, respectively. Therefore, the optimal layout can be defined as a set of points which belong to the half plane $\{(x,y)| \, x \geq 0\}$ that satisfy (4.40) and the optimality criteria $|\epsilon_1| = 1$. Following Rozvany *et al* [6, 8], the displacement field

$$u(x,y) = 0 \qquad \text{and} \quad v(x,y) = 2x \tag{4.41}$$

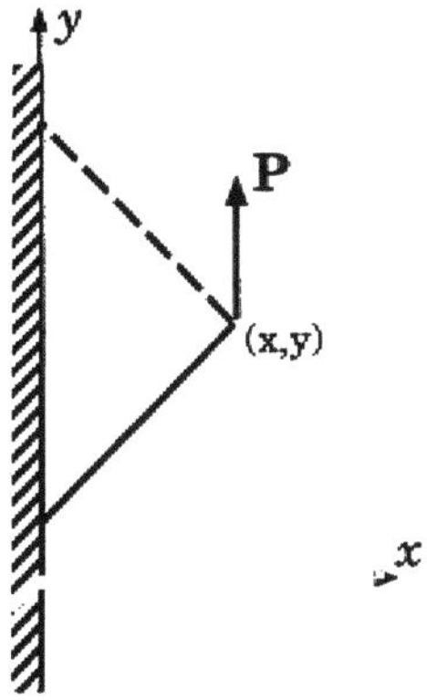

Figure 4.3 Illustrative example

satisfies (4.40) and leads to

$$\epsilon_x = du/dx = 0 \ , \quad \epsilon_y = dv/dy = 0 \ , \quad \gamma_{xy} = du/dy + dv/dx = 2 \ , \tag{4.42}$$

which results in

$$\epsilon_1 = 1 \ , \quad \epsilon_2 = -1 \ \text{ and } \ \alpha = \pm 45^\circ \tag{4.43}$$

where α is the direction of principal strains. The result obtained implies that the optimal bars must run at $\pm 45°$ to the vertical and must be in tension and compression, respectively in the two principal directions. By removing the potential members with zero forces, the two bar truss of Figure 4.3 is the optimum layout.

4.4 Mathematical model for the topological structural optimization problem

Let us consider the general linear elasticity problem subject to the applied body force $\mathbf{f}$ in a bounded open domain Ω and the surface traction forces $\mathbf{t}$ on Γ_t. Assume that Ω has a smooth boundary Γ comprising Γ_d, where displacements are prescribed and Γ_t, where traction forces are applied. It is also assumed that

$$\Gamma_t \cup \Gamma_d = \Gamma \qquad \text{and} \qquad \Gamma_t \cap \Gamma_d = \emptyset. \tag{4.44}$$

As described in Section 2.5, by applying the virtual displacement method, the equilibrium equation can be obtained by equating the external and internal virtual work. Assuming $\mathbf{u}$ to be the displacement field that defines equilibrium of the elastic structure and $\boldsymbol{v}$ to be the kinematically admissible virtual displacement field, so that

$$\boldsymbol{v} \in \mathbf{V} \quad where \quad \mathbf{V} = \left\{ \boldsymbol{v} | \ \boldsymbol{v} \in \left(H^1(\Omega) \right)^3 \text{ and } \boldsymbol{v} = \mathbf{0} \text{ on } \Gamma_d \right\}, \tag{4.45}$$

then for the elastic structure with a fixed boundary on Γ_d we have that

$$\int_{\Omega} \epsilon^T(v)(\mathbf{D}\epsilon(\mathbf{u}))d\Omega = \int_{\Omega} \mathbf{f}^T v d\Omega + \int_{\Gamma_t} \mathbf{t}^T v d\Gamma. \tag{4.46}$$

By borrowing the notation from functional analysis and using the energy bilinear form for the internal work and the load linear form for external work [15, 16], (4.46) can be written as

$$a(\mathbf{u}, v) = \ell(v) \ , \qquad \forall v \in \mathbf{V} \tag{4.47}$$

where

$$a(\mathbf{u}, v) = \int_{\Omega} \epsilon(v).(\mathbf{D}\epsilon(\mathbf{u}))d\Omega \tag{4.48}$$

and

$$\ell(v) = \int_{\Omega} \mathbf{f}.v d\Omega + \int_{\Gamma_t} \mathbf{t}.v d\Gamma. \tag{4.49}$$

The structural optimization problem considered in this book aims to find the stiffest possible structure using a certain amount of material. A structure with maximum global stiffness provides a minimum value for the external work with the real displacement field or minimum *mean compliance* $\ell(\mathbf{u})$. Therefore, the structural optimization problem with the mean compliance as the objective function can be constructed as,

$$\begin{array}{rl} \text{Minimize} & \ell(\mathbf{u}) \\ \text{subject to} & a(\mathbf{u}, v) = \ell(v) \quad \forall v \in \mathbf{V} \\ \text{and} & \textit{design restrictions} \end{array} \tag{4.50}$$

where design restrictions, $a(\mathbf{u}, v)$ and $\ell(v)$ are functions of the design variables. As we have seen before, the design variables of the structural topology optimization problem are the geometrical parameters representing the microscale perforations of the presumed material model (e.g. a, b and θ in the material model comprising square cells with rectangular holes or γ, μ and θ for the rank-2 material as explained in Chapter 3).

If the structure is fixed on the boundary Γ_d, then by substituting $\mathbf{u} \in \mathbf{V}$ instead of v in (4.47) the following equation is obtained

$$a(\mathbf{u}, \mathbf{u}) = \ell(\mathbf{u}). \tag{4.51}$$

Remembering that $\frac{1}{2}a(\mathbf{u}, \mathbf{u})$ is the strain energy, we observe that minimizing the mean compliance is equivalent to minimizing the strain energy.

On the other hand, the *total potential energy* can be written as

$$\Pi(\mathbf{u}) = \frac{1}{2}a(\mathbf{u}, \mathbf{u}) - \ell(\mathbf{u}) \tag{4.52}$$

and by substituting from (4.51) it follows that

$$\Pi(\mathbf{u}) = -\frac{1}{2}\ell(\mathbf{u}). \tag{4.53}$$

From (4.53) it is concluded that minimizing the mean compliance is equivalent to maximizing the total potential energy. Therefore, the optimization problem (4.50) can be written as

$$\begin{array}{rl} \text{Maximize} & \Pi(\mathbf{u}) \\ \text{subject to} & a(\mathbf{u}, v) = \ell(v) \quad \forall v \in \mathbf{V} \\ \text{and} & \textit{design restrictions}, \end{array} \tag{4.54}$$

whereas, by using the principal of minimum potential energy† (4.54) can be written as

$$\begin{array}{ll} \max\limits_{design} & \min\limits_{v \in \mathbf{V}} \ \Pi(v) \\ \text{subject to} & \textit{design restrictions}. \end{array} \tag{4.55}$$

Construction of the optimization problem based on the total potential energy is preferred, because it is more general and it always describes an optimization problem for the stiffest structure. While, in the case of prescribed boundary displacements, using the 'mean compliance' or the 'strain energy' as the objective function may lead to confusion [17]. This point can be realized by noticing the fact that in the case of having a non-zero prescribed displacement, and in the absence of body forces and tractions, the stiffest structure corresponds to the maximum strain energy (and maximum mean compliance), while in most of the cases the strain energy should be minimized.

In the general case of enforcing the prescribed boundary displacements $\bar{\mathbf{u}}$ on Γ_d, by using the exterior penalty method [2, 18] and applying a 'sufficiently large' penalty coefficient β, the total potential energy will be formed as

$$\Pi_\beta(v) = \int_\Omega \epsilon(v).\mathbf{D}\epsilon(\mathbf{u})d\Omega - \int_\Omega \mathbf{f}.v d\Omega - \int_{\Gamma_t} \mathbf{t}.v d\Gamma + \beta \int_{\Gamma_d} (v - \bar{\mathbf{u}})^2 d\Gamma. \tag{4.56}$$

It is noted that minimizing $\Pi_\beta(v)$ is equivalent to solving the equation of equilibrium and it yields the displacements $\mathbf{u}$.

If the material model with rectangular holes is used‡, considering a, b and θ as distributed functions and recalling the relations of Section 3.2.1, the optimization problem (4.55) can be expressed as

---

† According to the principal of minimum potential energy "among all the admissible displacement functions, the actual displacements make the total potential energy an absolute minimum [19]".

‡ In the case of rank-2 material model, because of the strong similarity it is enough to replace a and b with γ and μ.

$$\begin{array}{lll} \max\limits_{a,b,\theta} & \min\limits_{v\in\mathbf{V}} \ \Pi_\beta(v) & \\ \text{subject to} & \Omega_s = \int_\Omega (1-ab)d\Omega \leq \bar{\Omega}_s & (4.57) \\ \text{and} & 0 \leq a,b \leq 1 & \end{array}$$

where $\bar{\Omega}_s$ is the upper limit on the volume of the solid material. Equation (4.57) is equivalent to

$$\begin{array}{lll} \max\limits_{a,b,\theta} & \Pi(\mathbf{u}) & \\ \text{subject to} & \Omega_s \leq \bar{\Omega}_s & (4.58) \\ \text{and} & 0 \leq a,b \leq 1 & \end{array}$$

In the next section the optimality conditions for the adopted mathematical model (4.58), are derived.

4.5 Optimality criteria for the topological structural optimization problem

4.5.1 Optimality conditions

By using the Lagrange multipliers and the Kuhn-Tucker conditions, the necessary conditions for optimality in problem (4.58) are obtained. Adopting the finite element method for the analysis, we assume that the design variables a, b and θ, as explained in Section 3.2.1, are constant parameters inside each finite element. Therefore, (4.58) for a discretized problem can be written as

$$\begin{array}{llll} \text{Maximize} & \Pi(\mathbf{u}) & & \\ a^e, b^e, \theta^e & (e=1,\cdots,N) & & \\ \text{such that} & \sum_{e=1}^{N}(1-a^e b^e)\Omega^e - \bar{\Omega}_s \leq 0 & & \\ \text{and} & a^e - 1 \leq 0 & e=1,\cdots,N & (4.59) \\ & -a^e \leq 0 & e=1,\cdots,N & \\ & b^e - 1 \leq 0 & e=1,\cdots,N & \\ & -b^e \leq 0 & e=1,\cdots,N & \end{array}$$

where N is the number of elements and $\Pi(\mathbf{u})$ in the discretized domain is

$$\Pi(\mathbf{u}) = \frac{1}{2}\sum_{e=1}^{N}\int_{\Omega^e} \epsilon^T(\mathbf{u})\mathbf{D}^e\epsilon(\mathbf{u})d\Omega - \sum_{e=1}^{N}\int_{\Omega^e} \mathbf{u}^T\mathbf{f}d\Omega - \sum_{e=1}^{N}\int_{\Gamma_t^e} \mathbf{u}^T\mathbf{t}d\Gamma. \quad (4.60)$$

Now, introducing multipliers λ_{a0}^e, λ_{a1}^e, λ_{b0}^e, λ_{b1}^e, $(e = 1, \cdots, N)$ and Λ, the Lagrangian function $\mathcal{L}$ associated with problem (4.59) may be constructed as

$$\mathcal{L} = \Pi(\mathbf{u}) - \Lambda \left[\sum_{e=1}^{N} (1 - a^e b^e)\Omega^e - \bar{\Omega}_s \right] - \sum_{e=1}^{N} \lambda_{a0}^e(-a^e) - \sum_{e=1}^{N} \lambda_{a1}^e(a^e - 1)$$
$$- \sum_{e=1}^{N} \lambda_{b0}^e(-b^e) - \sum_{e=1}^{N} \lambda_{b1}^e(b^e - 1) \,. \qquad (4.61)$$

Stationarity of $\mathcal{L}$ with respect to a^e requires that

$$\frac{\partial \Pi(\mathbf{u})}{\partial a^e} + \Lambda b^e \Omega^e + \lambda_{a0}^e - \lambda_{a1}^e = 0\,, \qquad (e = 1, \cdots, N) \qquad (4.62)$$

and by substituting for $\Pi(\mathbf{u})$ from (4.60), it follows that

$$\frac{1}{2} \int_{\Omega^e} \boldsymbol{\epsilon}^T(\mathbf{u}) \frac{\partial \mathbf{D}^e}{\partial a^e} \boldsymbol{\epsilon}(\mathbf{u}) d\Omega - \int_{\Omega^e} \mathbf{u}^T \frac{\partial \mathbf{f}}{\partial a^e} d\Omega + \Lambda b^e \Omega^e + \lambda_{a0}^e - \lambda_{a1}^e = 0\,. \qquad (4.63)$$

Similarly stationarity of $\mathcal{L}$ with respect to b^e and θ^e yields

$$\frac{1}{2} \int_{\Omega^e} \boldsymbol{\epsilon}^T(\mathbf{u}) \frac{\partial \mathbf{D}^e}{\partial b^e} \boldsymbol{\epsilon}(\mathbf{u}) d\Omega - \int_{\Omega^e} \mathbf{u}^T \frac{\partial \mathbf{f}}{\partial b^e} d\Omega + \Lambda a^e \Omega^e + \lambda_{b0}^e - \lambda_{b1}^e = 0 \qquad (4.64)$$

and

$$\frac{1}{2} \int_{\Omega^e} \boldsymbol{\epsilon}^T(\mathbf{u}) \frac{\partial \mathbf{D}^e}{\partial \theta^e} \boldsymbol{\epsilon}(\mathbf{u}) d\Omega = 0\,, \qquad (e = 1, \cdots, N). \qquad (4.65)$$

Now, using the Kuhn-Tucker conditions, similar to (4.17), the switching conditions are obtained as

$$\sum_{e=1}^{N} (1 - a^e b^e)\Omega^e - \bar{\Omega}_s \leq 0\,; \quad \Lambda \left[\sum_{e=1}^{N} (1 - a^e b^e)\Omega^e - \bar{\Omega}_s \right] = 0\,; \quad \Lambda \geq 0\,, \qquad (4.66)$$

$$a^e - 1 \leq 0\,; \qquad \lambda_{a1}(a^e - 1) = 0\,; \qquad \lambda_{a1} \geq 0\,, \qquad e = 1, \cdots, N \qquad (4.67)$$

$$-a^e \leq 0\,; \qquad \lambda_{a0}(-a^e) = 0\,; \qquad \lambda_{a0} \geq 0\,, \qquad e = 1, \cdots, N \qquad (4.68)$$

$$b^e - 1 \leq 0\,; \qquad \lambda_{b1}(b^e - 1) = 0\,; \qquad \lambda_{b1} \geq 0\,, \qquad e = 1, \cdots, N \qquad (4.69)$$

$$-b^e \leq 0\,; \qquad \lambda_{b0}(-b^e) = 0\,; \qquad \lambda_{b0} \geq 0\,, \qquad e = 1, \cdots, N. \qquad (4.70)$$

The set of non-linear equations (4.63)-(4.70) provide $7N + 1$ equations (N is the number of finite elements) with the same number of unknowns to be solved.

The solution of these equations is a point of the design space which satisfies the necessary conditions of optimality. Since solving these equations is not an easy task, we try to establish an optimality criteria algorithm, instead. Based on these criteria an updating scheme is constructed. By starting from a feasible point in the design space and using the scheme, at each iteration the design variables are updated so that we gradually move towards the optimum point. A method based on this concept has proved to be computationally very efficient, although it can not be guaranteed that it always will converge to the global optimum.

4.5.2 Updating scheme

Employing the traditional optimality criteria method [20] an updating scheme for the design variables (e.g. a^e, b^e and θ^e) can be constructed. To derive this scheme, the mutual influences of the design variables on each other, also from element to element are ignored and in each iteration step they are updated independently.

Although the angle of rotation θ^e can be updated inside the iteration scheme, in order to increase the convergence speed, following Bendsøe and Kikuchi [17, 18] and making use of the results of research done on optimal orientation of orthotropic material, it can be dealt with separately. Therefore, in the updating scheme the cell size parameters a and b (γ and μ for rank-2 material) are resized. The issue of the optimal orientation is the subject of the Section 4.6.

If we define

$$E_a^e = \frac{\frac{1}{2}\int_{\Omega^e} \boldsymbol{\epsilon}^T(\mathbf{u})\,(\partial \mathbf{D}^e/\partial a^e)\,\boldsymbol{\epsilon}(\mathbf{u})d\Omega - \int_{\Omega^e} \mathbf{u}^T\,(\partial \mathbf{f}/\partial a^e)\,d\Omega}{-\Lambda b^e \Omega^e}\,, \tag{4.71}$$

then (4.63) can be written as

$$E_a^e = 1 + \frac{\lambda_{a0}^e}{\Lambda b^e \Omega^e} - \frac{\lambda_{a1}^e}{\Lambda b^e \Omega^e} \tag{4.72}$$

It is noted that if we assume that the lower and upper side constraints for a^e are not active (i.e. $0 < a^e < 1$), then $\lambda_{a0}^e = \lambda_{a1}^e = 0$ and (4.72) becomes: $E_a^e = 1$.

Now, let us imagine that in iteration k, the design variable a_k^e has been decreased in order to move towards optimum point. Therefore, $a_k^e < 1$ and the upper side limit is not active, which yields $\lambda_{a1}^e = 0$. Thus, noticing that $\Lambda b^e \Omega^e$ is a positive real number and $\lambda_{a0}^e \geq 0$, from (4.72) it follows that $E_a^e \leq 1$. On the other hand for increasing a_k^e we will get $E_a^e \leq 1$. So, inspired by this argument, we calculate the value of E_a^e and compare it with unity. If $E_a^e > 1$, then we let a_k^e decrease by the *move limit* ζ (i.e. $a_{k+1}^e = a_k^e(1-\zeta)$) and if $E_a^e < 1$, then $a_{k+1}^e = a_k^e(1+\zeta)$. Based on this conclusion and considering the side limits for a^e (i.e. $0 \leq a^e \leq 1$), Bendsøe *et al* [17] have suggested the following resizing scheme:

$$a_{k+1}^e =$$

$$\begin{cases} \min\{(1+\zeta)a_k^e, 1\} & \text{if} \quad a_k^e\,(E_a^e)_k \le \max\{(1-\zeta)a_k^e, 0\} \\ a_k^e\,[(E_a^e)_k]^\eta & \text{if} \quad \max\{(1-\zeta)a_k^e, 0\} < a_k^e\,(E_a^e)_k < \min\{(1+\zeta)a_k^e, 1\} \\ \max\{(1-\zeta)a_k^e, 0\} & \text{if} \quad \min\{(1+\zeta)a_k^e, 1\} \le a_k^e\,(E_a^e)_k \,. \end{cases} \tag{4.73}$$

We now define

$$E_b^e = \frac{\frac{1}{2}\int_{\Omega^e} \epsilon^T(\mathbf{u})\,(\partial \mathbf{D}^e/\partial b^e)\,\epsilon(\mathbf{u})d\Omega - \int_{\Omega^e} \mathbf{u}^T\,(\partial \mathbf{f}/\partial b^e)\,d\Omega}{-\Lambda a^e \Omega^e} \,. \tag{4.74}$$

Then, b^e can be updated as

$$b_{k+1}^e =$$

$$\begin{cases} \min\{(1+\zeta)b_k^e, 1\} & \text{if} \quad b_k^e\,\left(E_b^e\right)_k \le \max\{(1-\zeta)b_k^e, 0\} \\ b_k^e\,\left[\left(E_b^e\right)_k\right]^\eta & \text{if} \quad \max\{(1-\zeta)b_k^e, 0\} < b_k^e\,\left(E_b^e\right)_k < \min\{(1+\zeta)b_k^e, 1\} \\ \max\{(1-\zeta)b_k^e, 0\} & \text{if} \quad \min\{(1+\zeta)b_k^e, 1\} \le b_k^e\,\left(E_b^e\right)_k \,. \end{cases} \tag{4.75}$$

Here η is a damping factor and subscript k denotes the iteration number.

To have a monotonic algorithm, the approximate value of the damping factor η can be obtained by equating the resizing values for the switching points ($E = 1 - \zeta$ and $E = 1 + \zeta$) which leads

$$(1-\zeta)^\eta = 1 + \zeta \tag{4.76}$$

and

$$(1+\zeta)^\eta = 1 - \zeta. \tag{4.77}$$

From (4.76) and (4.77) it respectively follows that

$$\eta = \frac{\ln(1+\zeta)}{\ln(1-\zeta)} \tag{4.78}$$

and

$$\eta = \frac{\ln(1-\zeta)}{\ln(1+\zeta)} \,. \tag{4.79}$$

Assuming ζ to be a small value and using the Taylor series expansion, for both (4.78) and (4.79) it follows that

$$\eta \simeq -1 + \zeta \,. \tag{4.80}$$

It is important to note that $(E_a^e)_k$ and $(E_b^e)_k$ depend on the current value of Λ_k, the Lagrange multiplier of the active volume constraint, which needs to be adjusted in an inner iteration loop. For this purpose the standard methods like the Newton-Raphson iterative procedure or the bisection method can be employed [18]. As the Newton-Raphson method requires the extra computational effort of calculating $d\Omega_s/d\Lambda$ (e.g. by the finite difference method), here the bisection method is adopted. According to this procedure, and noting that for the static loading problem $d\Omega_s/d\Lambda < 0$, the steps below are followed:

1. Find Λ_{min}^0 and Λ_{max}^0 such that
$$\Omega_s\left(\Lambda_{min}^0\right) > \bar{\Omega}_s \qquad \text{and} \qquad \Omega_s\left(\Lambda_{\max}^0\right) < \bar{\Omega}_s \ .$$
2. Calculate
$$\Lambda^m = \frac{1}{2}\left(\Lambda_{min}^m + \Lambda_{max}^m\right) \ .$$
3. Calculate $\Omega_s\left(\Lambda^m\right)$ and update either Λ_{min} or Λ_{max} as below
$$\begin{cases} \text{if} \ \ \Omega_s < \bar{\Omega}_s \ \ \text{then} \ \ \Lambda_{max}^{m+1} = \Lambda^m \\ \\ \text{if} \ \ \Omega_s > \bar{\Omega}_s \ \ \text{then} \ \ \Lambda_{min}^{m+1} = \Lambda^m \end{cases}$$
4. Increase the iteration number by one, and repeat steps 2 and 3 until $|\Omega_s - \bar{\Omega}_s| \leq \delta$.

Here, δ indicates the acceptable tolerance for the volume constraint.

Note that for stability of the algorithm, the move size ζ and the tolerance for the volume constraint δ have to be adjusted. As a general rule, in order to have a larger move size ζ, the volume constraint must be more relaxed. This issue will be dealt with later is Chapter 5.

4.5.3 A modified resizing scheme

In the resizing scheme which was described, it is noted that the difference between the values of the cell size parameters (design variables) in two successive iteration (e.g. a_k^e and a_{k+1}^e) is proportional to the magnitude of the parameters. This means that, while the larger values of a^e and b^e vary rapidly, the updating rates for small values are very small. In fact, an infinite number of iterations are needed to reach zero. Use of conditional 'if' commands in the scheme may affect the stability of the algorithm and they are better avoided.

In order to have a better monotonic and stable algorithm, by following an argument similar to the last section, the following resizing scheme is suggested:

$$a_{k+1}^e =$$

$$\begin{cases} \min\{(1+\frac{\zeta}{|a_k^e-\zeta|})a_k^e, 1\} & \text{if} \quad a_k^e\,(E_a^e)_k \le \max\{(1-\zeta)a_k^e, 0\} \\ a_k^e\,[(E_a^e)_k]^{-\frac{1}{a_k^e}} & \text{if} \quad \max\{(1-\zeta)a_k^e, 0\} < a_k^e\,(E_a^e)_k < \min\{(1+\zeta)a_k^e, 1\} \\ \max\{(1-\frac{\zeta}{|a_k^e-\zeta|})a_k^e, 0\} & \text{if} \quad \min\{(1+\zeta)a_k^e, 1\} \le a_k^e\,(E_a^e)_k \end{cases} \tag{4.81}$$

and

$$b_{k+1}^e =$$

$$\begin{cases} \min\{(1+\frac{\zeta}{|b_k^e-\zeta|})b_k^e, 1\} & \text{if} \quad b_k^e\,(E_b^e)_k \le \max\{(1-\zeta)b_k^e, 0\} \\ b_k^e\,[(E_b^e)_k]^{-\frac{1}{b_k^e}} & \text{if} \quad \max\{(1-\zeta)b_k^e, 0\} < b_k^e\,(E_b^e)_k < \min\{(1+\zeta)b_k^e, 1\} \\ \max\{(1-\frac{\zeta}{|b_k^e-\zeta|})b_k^e, 0\} & \text{if} \quad \min\{(1+\zeta)b_k^e, 1\} \le b_k^e\,(E_b^e)_k \end{cases} \tag{4.82}$$

With a similar move size ζ this modified scheme is much faster than the former one.

4.6 Optimal orientation

As it noted earlier, the design variables of the structural topology optimization problem consist of sizing variables (a^e and b^e) and orientation variables (θ^e). In the last section the resizing scheme for the size variables was considered. In this section the issue of optimal values for the orientation variables are discussed. Having determined the sizing design variables a^e and b^e by resizing scheme, the current problem is how to determine θ^e to maximize (or minimize) the objective function.

By means of the coordinate transformation formulae, we are able to relate the elasticity properties of the fixed system, $\mathbf{D}^H(a^e, b^e)$, and the rotated system, $\mathbf{D}^G(a^e, b^e, \theta^e)$. The general elasticity tensor $E_{ijk\ell}^G$ can be computed by [21]

$$E_{ijk\ell}^G(a^e, b^e, \theta^e) = \sum_{p=1}^{2}\sum_{q=1}^{2}\sum_{r=1}^{2}\sum_{s=1}^{2} a_{ip}(\theta^e)a_{jq}(\theta^e)a_{kr}(\theta^e)a_{\ell s}(\theta^e)E_{pqrs}^H(a^e, b^e) \tag{4.83}$$

where in the case of 2-D problems, the rotation matrix is defined by

$$\mathbf{a}(\theta^e) = \begin{bmatrix} \cos\theta^e & -\sin\theta^e \\ \sin\theta^e & \cos\theta^e \end{bmatrix} . \tag{4.84}$$

Note that $\mathbf{D}^G$ is the compact form of E^G_{ijkl} (see Appendix A). We also note that the relation (3.4) can be used. That is

$$\mathbf{D}^G(a^e, b^e, \theta^e) = \mathbf{R}^T(\theta^e)\mathbf{D}^H(a^e, b^e)\mathbf{R}(\theta^e), \tag{4.85}$$

where by equating the strain energy density of the fixed and rotated frames, $\mathbf{R}(\theta^e)$ is obtained as

$$\mathbf{R}(\theta^e) = \begin{bmatrix} \cos^2\theta^e & \sin^2\theta^e & \cos\theta^e \sin\theta^e \\ \sin^2\theta^e & \cos^2\theta^e & -\cos\theta^e \sin\theta^e \\ -2\cos\theta^e \sin\theta^e & 2\cos\theta^e \sin\theta^e & \cos^2\theta^e - \sin^2\theta^e \end{bmatrix}. \tag{4.86}$$

Now, we consider the problem of how to update the design variables. Apart from mathematical programming methods, the following methods have been used by researchers to update the orientational design variable θ^e:

1. principal strain method (Pederson's method),
2. principal stress method,
3. stress based optimality criteria method, and
4. strain based optimality criteria method.

1. Principal strain method: This method is based on Pederson's discussion [22] on the problem of optimal orientation of an orthotropic material in order to obtain an extreme strain energy density. Noting (4.51), the strain energy density is equal to one half of the mean compliance in a unit volume. In Pederson's problem, the elasticity matrix $\mathbf{D}$ is fixed and a rotating strain field exists. Writing the strain energy density in terms of principal strains, he concluded that "material axes which are also principal strain axes always give stationary energy density" [22].

Cheng *et al* [23] have argued that this consideration does not fulfil the requirements of the real design problem and the results of this method are relatively poor. The discussion by Pederson is local and is limited to only a special point of the design domain. Hence, it is applicable to a unit cell case where the orientational design variable is separated from the whole design domain to obtain extreme strain energy density. In other words, instead of stationarity of the strain energy density, the strain energy corresponding to the whole design domain has to be considered.

In the topology design problem, the strain field and principal strains and their directions are dependent on the orientation variable and can change when the orientation variable θ is varied. In fact, these changes can be caused not only by the variations of θ at the considered point, but also the variations of the other spatial points of the design domain. Especially when the finite element discretization is used, the coupling will exist between the discrete orientation variables of the various finite elements. This method was suggested by Bendsøe [24] as an alternative to an iterative Newton type algorithm. "For materials with square microvoids, the cells should be rotated along the directions of principal strain" [24].

2. Principal stress method: The use of the principal stress directions as the optimal orientation was considered by Suzuki and Kikuchi [18]. The intuitive idea behind this is that, having a constant stress field with respect to the orientational variable, to obtain the stiffest structure, the principal material axis should be coalescent with the principal stress direction. This method, although lacking a mathematical basis, provides quite good results (c.f. [23]).

Díaz and Bendsøe [25] extended this method for multiple load cases. In their approach, first the principal stress directions for different load cases are determined and then by considering the obtained equation from stationarity of the Lagrangian with respect to θ^e, a combined equation is obtained. By solving this equation the optimal orientation is determined.

3. Strain based optimality criteria method: Considering (3.4), the sensitivity of $\mathbf{D}(a^e, b^e, \theta^e)$ with respect to the orientational variation in element e can be written as

$$\frac{\partial \mathbf{D}^e}{\partial \theta^e} = \frac{\partial \mathbf{R}^T(\theta^e)}{\partial \theta^e}\mathbf{D}^H(a^e, b^e)\mathbf{R}(\theta^e) + \mathbf{R}^T(\theta^e)\mathbf{D}^H(a^e, b^e)\frac{\partial \mathbf{R}(\theta^e)}{\partial \theta^e}. \tag{4.87}$$

Substituting for $\mathbf{R}(\theta^e)$ from (4.86) in (4.87) and applying in (4.65) results in

$$\alpha_1^e \cos 2\theta^e - \alpha_2^e \sin 2\theta^e + \alpha_3^e \cos 4\theta^e - \alpha_4^e \sin 4\theta^e = 0, \quad (e = 1, \cdots, N) \tag{4.88}$$

where

$$\alpha_1^e = \frac{1}{2}(D_{11}^H - D_{22}^H) \int_{\Omega^e} \epsilon_{12}(\epsilon_{11} + \epsilon_{22}) d\Omega \tag{4.89a}$$

$$\alpha_2^e = \frac{1}{2}(D_{11}^H - D_{22}^H) \int_{\Omega^e} (\epsilon_{11}^2 - \epsilon_{22}^2) d\Omega \tag{4.89b}$$

$$\alpha_3^e = \frac{1}{2}(D_{11}^H + D_{22}^H - 2D_{12}^H - 4D_{66}^H) \int_{\Omega^e} \epsilon_{12}(\epsilon_{11} - \epsilon_{22}) d\Omega \tag{4.89c}$$

$$\alpha_4^e = \frac{1}{4}(D_{11}^H + D_{22}^H - 2D_{12}^H - 4D_{66}^H) \int_{\Omega^e} \left[(\epsilon_{11} - \epsilon_{22})^2 - \epsilon_{12}^2\right] d\Omega. \tag{4.89d}$$

We observe that equations (4.88) are N nonlinear coupled equations which should be solved by iterative techniques. If we assume that the coefficients are not sensitive to the variations of the discretized orientational variable θ^e, then (4.89) can be approximately solved for each individual element.

4. Stress based optimality criteria method: This procedure is very similar to the strain based method, but instead of strains, stresses are considered. Substituting $\boldsymbol{\epsilon} = \mathbf{D}^{-1}\boldsymbol{\sigma} = \mathbf{C}\boldsymbol{\sigma}$ in (4.65) it yields

$$\int_{\Omega^e} \boldsymbol{\sigma}^T \frac{\partial \mathbf{C}^e}{\partial \theta^e} \boldsymbol{\sigma} d\Omega = 0, \qquad (e = 1, \cdots, N). \tag{4.90}$$

Following a procedure similar to the strain based method results in the equations below which are very similar to the strain based case:

$$\beta_1^e \cos 2\theta^e - \beta_2^e \sin 2\theta^e + \beta_3^e \cos 4\theta^e - \beta_4^e \sin 4\theta^e = 0, \quad (e = 1, \cdots, N) \quad (4.91)$$

where

$$\beta_1^e = \frac{1}{2}(C_{11}^H - C_{22}^H) \int_{\Omega^e} \sigma_{12}(\sigma_{11} + \sigma_{22}) d\Omega, \quad (4.92a)$$

$$\beta_2^e = \frac{1}{2}(C_{11}^H - C_{22}^H) \int_{\Omega^e} (\sigma_{11}^2 - \sigma_{22}^2) d\Omega, \quad (4.92b)$$

$$\beta_3^e = \frac{1}{2}(C_{11}^H + C_{22}^H - 2C_{12}^H - 4C_{66}^H) \int_{\Omega^e} \sigma_{12}(\sigma_{11} - \sigma_{22}) d\Omega, \quad (4.92c)$$

$$\beta_4^e = \frac{1}{4}(C_{11}^H + C_{22}^H - 2C_{12}^H - 4C_{66}^H) \int_{\Omega^e} \left[(\sigma_{11} - \sigma_{22})^2 - \sigma_{12}^2\right] d\Omega. \quad (4.92d)$$

As with the last case, the set of equations (4.91) may be solved by iterative solution techniques.

Cheng *et al* [23] in their research have concluded that "strain response is much more sensitive than the stress response for both statically determinate and indeterminate structures and much stronger coupling exists among the orientational variables when the strain field is used in comparison with the stress field". So use of stress based methods is preferred to the strain based methods. Furthermore, especially for single objective function optimization problems, the results obtained by the principal stress method are very close to the general stress based optimality criteria method.

In the enclosed software an approach based on the principal stress method is adopted. In each iteration, θ^e is updated according to the angle of the principal direction at the centre of the element. For example, in the case of quadrilateral elements, if 3×3 Gauss points are used, the principal direction obtained for the middle Gauss point can be used. If 2×2 Gauss points are used, the average of the angles obtained for the Gauss points can be used as the updated value of θ^e in each iteration.

4.7 Algorithm

Before explaining the algorithm let us summarize what we have achieved so far. The structural topology optimization by the homogenization method comprises two different modules: homogenization and optimization.

In homogenization module our aim is to establish an elastic constitutive relationship as a function of the size parameters of cellular microstructures that comprise the presumed material of each element. This is done only once and stored for all time. For rank-layered material model, this is achieved by analytical solution of the homogenization equations. For the material model with rectangular holes, we solved the homogenization equation for a set of values of a and b and then by least square fitting we expressed the elements of the homogenized elasticity matrix as an explicit polynomial in terms of values of a and b. Therefore having the values of a^e and b^e in each element the elasticity matrix $\mathbf{D}(a^e, b^e)$ is determined. As an easy alternative the artificial material may be used.

It was also shown that by using frame transformation formulae for all material models, how the effects of the orientational variables θ^e were also considered. In summary by use of the homogenization module $\mathbf{D}(a^e, b^e, \theta^e)$ is obtained as an explicit function of its arguments. In the case of rank-2 materials a^e and b^e are replaced by γ^e and μ^e, respectively. It is noted that calculation of the sensitivities of $\mathbf{D}$, which are required in the updating scheme is straightforward.

In the optimization module, considering the geometrical parameters of the presumed material model in finite elements as design variables, the total potential energy was adopted as the objective function. The volume of material was considered as the global constraint which has to be active. Forming the optimization problem, by the optimality criteria method, an updating scheme was constructed.

Now, we are in the position to construct the optimization algorithm:

1. Define the reference domain and decide upon the design and non-design domains. The design domain is the part of the domain where the microstructure is allowed to vary in the optimization process. The material of the non-design domain is fixed and is usually solid material or completely void. Choosing a suitable reference domain that allows us to introduce surface tractions and boundaries is often accomplished by intuition and engineering judgement. Here the loads, boundary conditions and the volume of material are also defined.
2. Discretize the reference domain by generating a structured or unstructured finite element mesh for analysis.
3. Initialize the design variables (i.e. the values of a^e, b^e and θ^e). It should be noted that initially a^e and b^e may have different values for different elements but they have to provide a material volume close to the given volume of material. The starting values for θ^e may be zero.
4. By means of the homogenization module, construct the elasticity matrices for each element $\mathbf{D}(a^e, b^e, \theta^e)$. Explicit derivatives of $\mathbf{D}$ with respect to the design variables are also calculated.
5. Carry out a finite element analysis using the above constitutive matrices for each element and obtain displacements.
6. Evaluate the objective function and the mutual energies. Here E_a^e and E_b^e are computed according to (4.71) and (4.74).
7. Use a resizing scheme to update the design variables. Note that the scheme is affected by the value of the Lagrange multiplier of the volume constraint

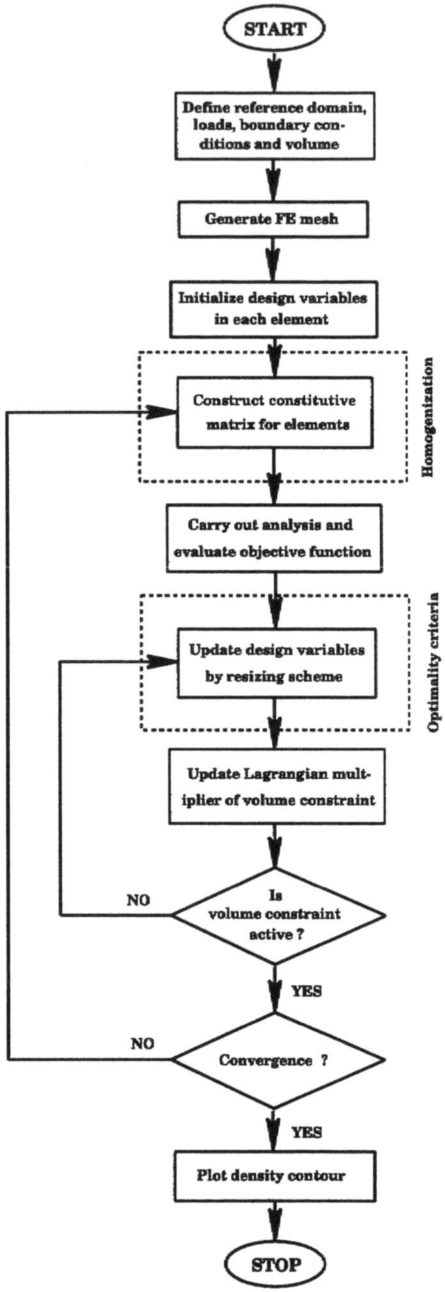

Figure 4.4 *Algorithm for topological structural optimization using homogenization and optimality criteria methods.*

(Λ in (4.61)) and consequently it has to be adjusted.

8. Check if the volume constraint is active, if it is not, update it and go back to step 7, otherwise continue.
9. Produce a new design based on the new set of updated design variables (a^e, b^e, θ^e) for each element e.
10. Check if solution has converged, if not then use the updated design variables and go back to step 4, otherwise continue.
11. Produce a density contour plot to establish optimal topology. Other post-processing techniques (e.g. image processing) may be used to obtain a clearer image with smoother boundaries.

The above algorithm is illustrated in Figure 4.4 . The convergence criteria may be constructed based on the rate of the objective function or the design variables. For example, the program can be terminated when in two (or a few) successive iterations the change in any of the design variables is not more than a given value and/or the rate of the improvement of the objective function is less than a certain value.

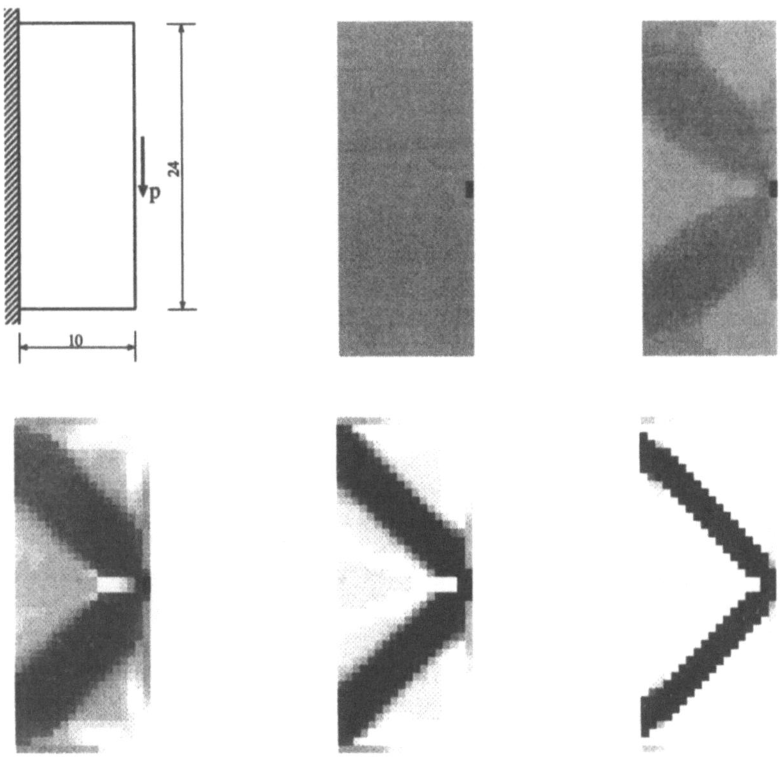

Figure 4.5 Iteration history of Example1.

4.8 Examples

Based on the algorithm described in the last section, the enclosed FORTRAN program given the name PLATO (PLAne Topology Optimization) has been developed. This program is based on the optimality criteria method and allows the user to have the option to choose different material models, different element types and different resizing schemes. Three examples are now presented to illustrate the use of PLATO.

Example 1: This example involves a clamped deep beam with a single load at the centre of the right edge with $E = 100000.$, $\nu = 0.3$ and the required solid fraction $V_{\text{solid}}/V = 20\%$. All units are assumed to be consistent.

The analytical solution of this problem was discussed in Section 4.3.2. It was shown that the optimal layout is a two bar truss (see Figure 4.3). In this example a mesh of 8-node elements is adopted and for the material model, cells with rectangular voids are used. Initially the material is distributed uniformly. The problem definition, reference domain and a few intermediate and the final result are illustrated in Figure 4.5.

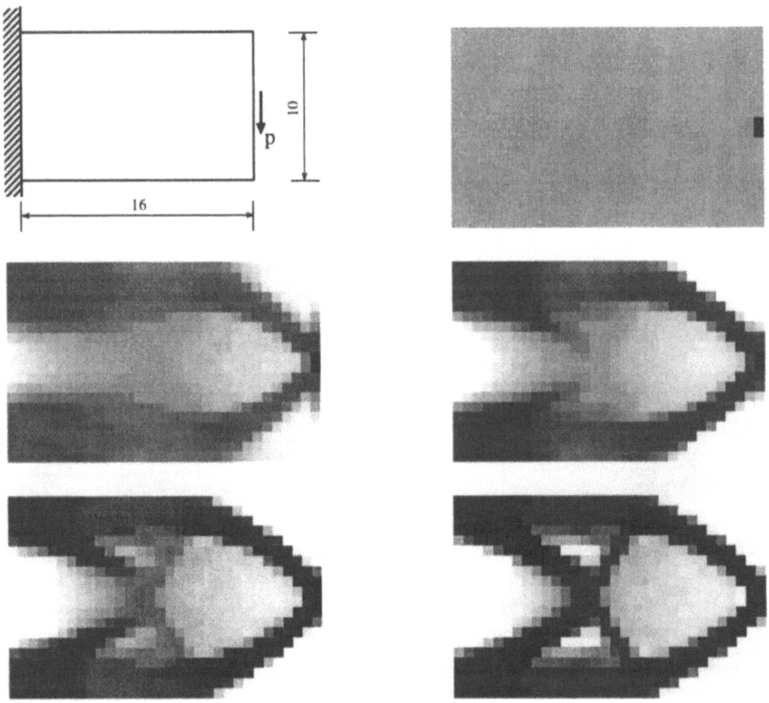

Figure 4.6 Iteration history of the clamped beam of Example 2.

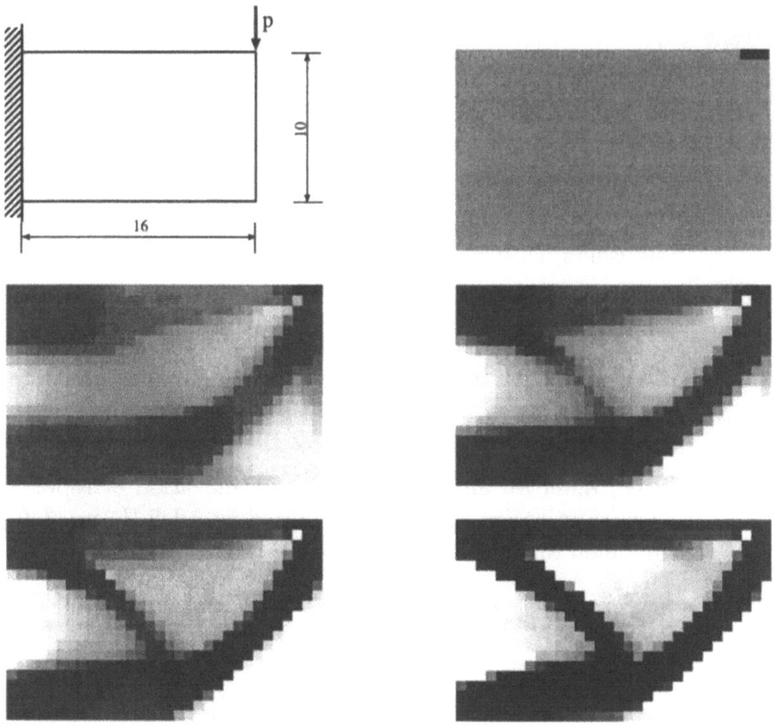

Figure 4.7 *Iteration history of the beam of Example 3.*

Example 2: In the next example, a clamped beam with a single load at the centre of the right edge is considered. $E = 100000.$, $\nu = 0.3$ and $V_{\text{solid}}/V = 50\%$. The problem definition and design and non-design domains are respectively illustrated by the top left and the top right pictures of Figure 4.6. The same element type and material model as in Example 1 is used and results in the shape given in Figure 4.6.

Example 3: This example is the same as Example 2, but the point load is applied to the top right corner. The problem definition, iteration history and the final result is illustrated in Figure 4.7.

The next chapter is devoted to experiences in topology optimization of plane problems in order to study the influence of the main parameters on the optimal topology.

References

[1] Haftka R.T. and Gürdal Z., *Elements of structural optimization.* Kluwer Academic Publishers, Dordrecht, third edn. (1992)

[2] Kirsch U., *Structural optimization.* Springer-Verlag, Berlin (1993)

[3] Rozvany G.I.N. and Zhou M., The COC algorithm, Part I: Cross section optimization or sizing, *Comp. Meth. in Appl. Mech.*, **89**, 281–308 (1991)

[4] Michell A.G.M, The limits of economy of material in frame-structures, *Philosophical Magazine*, **8**, 305–316 (1904)

[5] Rozvany G.I.N., *Structural design via optimality criteria.* Kluwer Academic Publishers, Dordrecht (1989)

[6] Rozvany G.I.N., Bendsøe M.P. and Kirsch U., Layout optimization of structures, *Appl. Mech. Rev.*, **48**(2), 41–119 (1995)

[7] Rozvany G.I.N. and Zhou M., Optimality criteria methods for large discretized systems, in *Advaces in design optimization*, edited by Adeli H., pp. 41–108. Chapman & Hall, London (1994)

[8] Rozvany G.I.N., Zhou M. and Sigmund O., Optimization of topology, in *Advaces in design optimization*, edited by Adeli H., pp. 340–399. Chapman & Hall, London (1994)

[9] Zhou M. and Rozvany G.I.N., DCOC: an optimality criteria method for large systems. Part I: theory, *Structural Optimization*, **5**, 12–25 (1992)

[10] Rozvany G.I.N. (ed.), *Shape and layout optimization of structural systems and optimality criteria methods.* Springer-Verlag, CISM, Udine (1992)

[11] Arora J.S., Huang M.W. and Hsieh C.C., Methods for optimization of nonlinear problems with discrete variables: a review, *Structural Optimization*, **8**, 69–85 (1994)

[12] Kuhn H.W. and Tucker A.W., Nonlinear programming, in *Proceedings of the second Berkeley symposium on mathematical statistics and probability*, edited by Neyman J., pp. 481–492 (1951)

[13] Wismer D.A. and Chattergy R., *Introduction to nonlinear optimization.* North Holland, Newyork (1979)

[14] Olhoff N. and Taylor J.E., On structural optimization, *J. Appl. Mech.*, **50**, 1139–1151 (1983)

[15] Reddy J.N. and Rasmussen M.L., *Advanced engineering analysis.* John Wiley & Sons, New York (1982)

[16] Reddy J.N., *Applied functional analysis and variational methods in engineering.* McGraw-Hill, New York (1986)

[17] Bendsøe M.P., Díaz A.R. and Kikuchi N., Topology and generalized layout optimization of elastic structures, in *Topology design of structures*, edited by Bendsøe M.P. and Mota Soares C. A., pp. 159–205. Kluwer Academic Publishers (1993)

[18] Suzuki K. and Kikuchi N., A homogenization method for shape and topology optimization, *Comp. Meth. Appl. Mech. Eng.*, **93**, 291–318 (1991)

[19] Washizu K., *Variational methods in elasticity and plasticity.* Pergamon Press, Oxford, third edn. (1982)

[20] Berke L. and Khot N.S., Use of optimality criteria for large scale systems, *AGARD LS-70* (1974)

[21] Chung T.J., *Continuum mechanics.* Prentice-Hall, London (1988)

[22] Pedersen P., On optimal orientation of orthotropic materials, *Structural Optimization*, **1**, 101–106 (1989)

[23] Cheng H.C., Kikuchi N. and Ma Z.A., An improved appproach for determining the optimal orientation of orthotropic material, *Structural Optimization*, **8**, 101–112 (1994)

[24] Bendsøe M.P., Optimal shape design as a material distribution problem, *Structural Optimization*, **1**, 193–202 (1989)

[25] Díaz A.R. and Bendsøe M.P., Shape optimization of multipurpose structures by a homogenization method, *Structural Optimization*, **4**, 17–22 (1992)

5

EXPERIENCES IN TOPOLOGY OPTIMIZATION OF PLANE STRESS PROBLEMS

The outcome of the topology optimization process is influenced by several factors: different material models, resizing schemes, finite element discretizations and element types. The optimal layout may also be affected by the resizing parameters. This chapter is mainly devoted to the study of the influence of the above parameters. To demonstrate the usefulness of the method, several examples of a more practical nature are provided.

5.1 Introduction

In Chapter 4 the PLATO code was introduced and three examples were presented to illustrated the procedure. In this chapter, by means of further examples, the different influences that various factors can have on the optimization process will be highlighted. Such information will be invaluable to the structural design engineer when using topology optimization and when employed together with good engineering intuition will provide a robust and powerful design strategy.

As pointed out in the last chapter, the first task in topological structural optimization is to define the reference domain and then divide it into design and non-design sub-domains. This decision is usually based on the structural functionality, geometrical limitations and the experience of designer. The non-design domain may comprise solid or void regions. It should be noted that except for any holes, the whole domain is discretized by finite elements. As an example the design and non-design domains and the finite element mesh of Example 1 of Chapter 4 is depicted in Figure 5.1 where the two middle elements adjacent to the load P are considered as a solid, non-design domain.

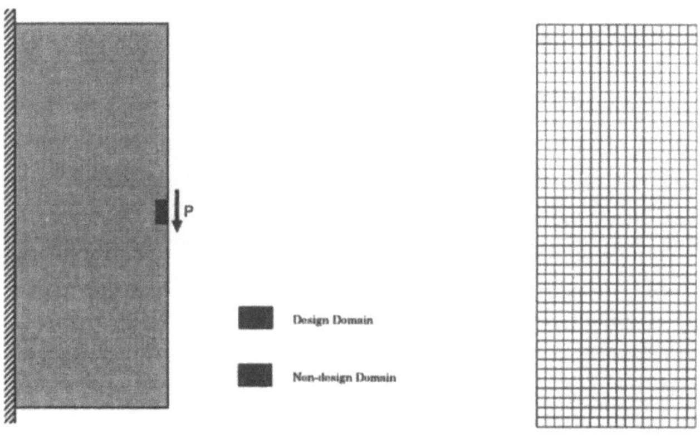

Figure 5.1 *Design and non-design domains (left) and the finite element mesh (right).*

The program PLATO initializes the design domain by distributing the available material evenly to all elements. However, it is possible to start with a different material distribution by reading the related data from a file.

To study the influence of different factors, the problem of the *short cantilever beam* (SCB) shown in Figure 5.2 will mainly be used. Experiments on this example have been carried out by several researchers. See for example Bendsøe *et al* [1] and Suzuki *et al* [2]. In the following some more examples are presented to demonstrate the usefulness of the method.

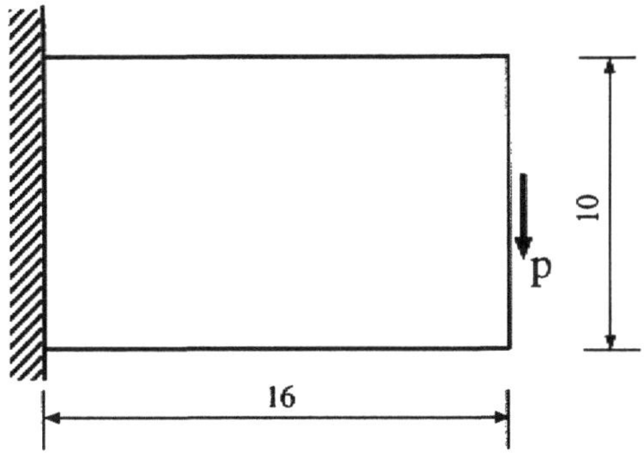

Figure 5.2 *Problem definition of short cantilever beam (SCB).*

5.2 Effect of material model

As explained earlier in Chapter 3, different material models can be used in structural topology optimization problems. The material model of square microcells with rectangular holes, the rank-2 material model and the artificial material model are all available in the PLATO code. For the material model with rectangular holes, two different least-squares complete polynomials are provided: bi-cubic and bi-quartic. See Section 3.4.5.

The geometry, loading and boundary conditions of the SCB problem are illustrated in Figure 5.2. A point load of intensity $P = 1000$ is applied midway down the right hand side of the beam. Unless otherwise stated, in all of the following examples the following problem parameters are adopted: modulus of elasticity $E = 100000$, Poisson's ratio, $\nu = 0.3$ and volume fraction $V_{\text{solid}}/V = 40\%$. All units are assumed to be consistent. Unless otherwise stated 8-noded isoparametric elements are used with 3×3 Gauss-Legendre quadrature.

5.2.1 Material model with rectangular holes

In the first experiment the material model of square microcells with rectangular voids with bi-cubic interpolation is tried. The domain of the problem is divided into a mesh of 32×20 equal elements. The two middle elements adjacent to the point load are considered as non-design elements. The move limit ζ and the volume tolerance δ are taken as 0.008 and 0.001 respectively†. The

† In practice far larger values are usually used.

modified resizing scheme of Section 4.5.3 is employed and the layout obtained after 130 iterations is shown in Figure 5.3.

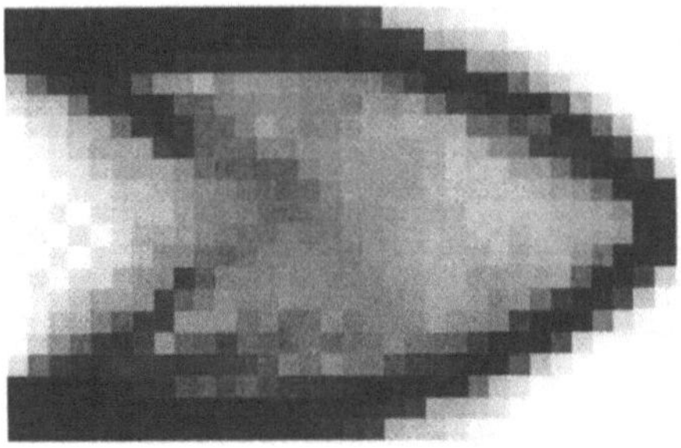

Figure 5.3 *Optimal layout using the material model with rectangular holes and bi-cubic polynomial.*

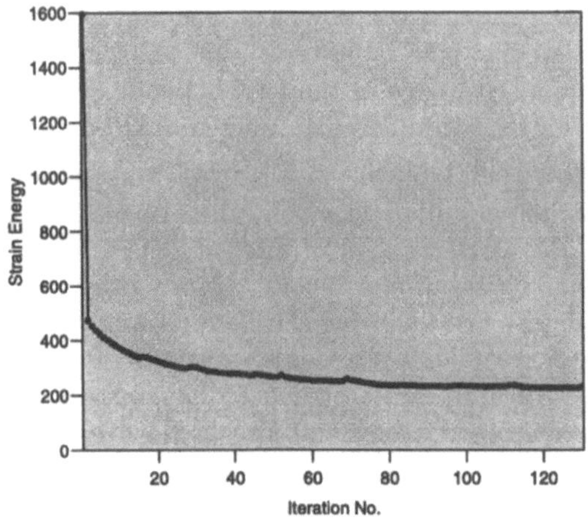

Figure 5.4 *Iteration history for the material model with rectangular holes and a bi-cubic least squares polynomial.*

The iteration history is illustrated in Figure 5.4. The strain energy (or mean compliance) of the initial structure is 1594.2 and after the first iteration it drops to 472.8. The reason for this dramatic drop in the value of strain energy is the alignment of the major axis of the material with the maximum principal stress direction. After 130 iterations the optimal strain energy is equal to 226.3. In the next example a bi-quartic polynomial is used instead of as bi-cubic one for the least squares representation of the data obtained from the solution of the homogenization equations. The layout of Figure 5.5 results when PLATO is executed for the SCB problem with this material model while the other parameters are exactly as before. The iteration history is illustrated in Figure 5.6. The optimum value of the strain energy is equal to 227.0.

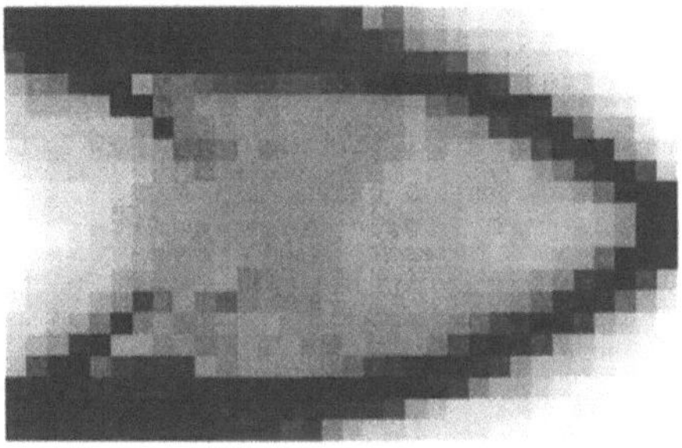

Figure 5.5 *Optimal layout using the material model with rectangular holes and bi-quartic polynomial.*

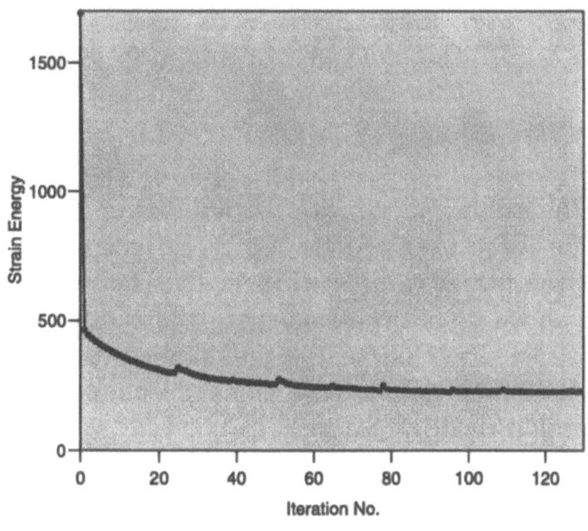

Figure 5.6 *Iteration history for the material model with rectangular holes and a bi-quartic least squares polynomial.*

An almost identical result to that of the bi-cubic case is obtained. The iteration history is very similar to the bi-cubic case, but the starting strain energy is slightly higher (1690.8) than for the bi-cubic case.

5.2.2 Artificial material model

The SCB problem is now solved with the artificial material model. The move limit, the volume tolerance, the finite element mesh and other parameters are chosen to be identical to those given in Section 5.2.1. The penalty exponent of the artificial material given in (3.20) is chosen as $\mu = 4.0$. Using the modified resizing scheme, the layout of Figure 5.7 is obtained. As can be seen the layout obtained has fewer grey areas and the structural form is quite clear. This is desirable from a practical point of view where construction of the porous material with infinitesimal holes is rather impractical and expensive.

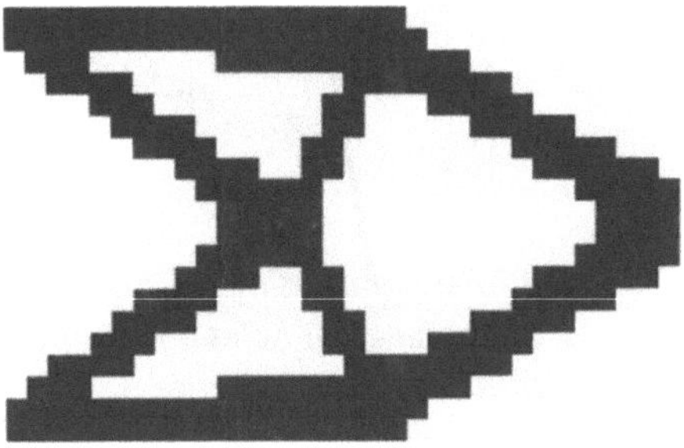

Figure 5.7 *Optimum layout obtained by using artificial material model with* $\mu = 4.0$.

The variation of the strain energy with the number of iterations is illustrated in Figure 5.8. The value obtained for the strain energy after 130 iterations is 255.9, whereas the initial value is 4219.1. It is interesting to note that in contrast with the above cases the strain energy has not changed after aligning the principal material direction to the maximum principal stress direction. The reason is that the artificial material model used here is in fact an isotropic material. See equation (3.20).

It is important to note that *the value obtained for the strain energy of the structure when the artificial material model is used should not be directly compared with the values obtained from the other material models.* The reason for this is that the elasticity matrix of the artificial material is merely a convenient fiction.

One way to find a more useful strain energy value of the obtained layout is to re-interpret the material as though it were a microcellular type material with square microcells and rectangular holes and use the *homogenization technique* to re-evaluate the equivalent strain energy. Thus, we find the parameters (here a and b which in the case of isotropic artificial material are always equal) for the elements of the optimal layout and then use the results obtained from the

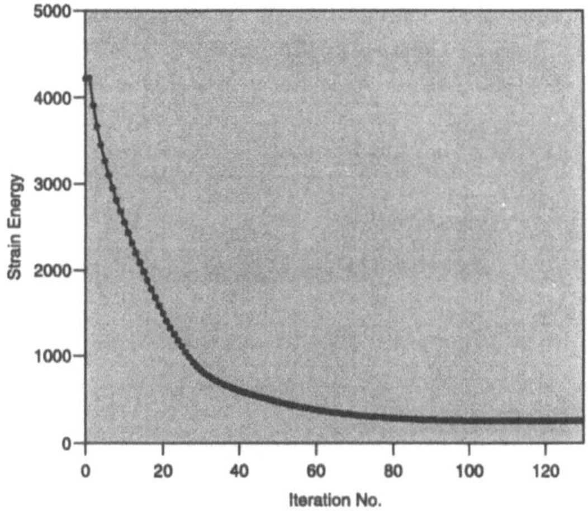

Figure 5.8 *Iteration history for the artificial material model with* $\mu = 4.0$.

homogenization theory to find the equivalent elasticity coefficients. Finally, we carry out a subsequent finite element analysis and calculate the real strain energy.

For example, if the bi-cubic least squares polynomial obtained for the material model with rectangular holes is used, then the strain energy value after 130 iterations will be equal to 243.3 instead of 255.9. It is also observed that when compared to the material model with rectangular voids this value (243.3) is 7.5% higher. In fact, this extra value for the strain energy is the penalty that we pay to suppress the porous (grey) regions by penalizing them. The effect of the penalty power μ which is responsible for this, will be studied below.

As expected, in Figure 5.7 a highly symmetric result is obtained. This would seem to indicate that the optimization algorithm of PLATO is not responsible for the small asymmetries of Figures 5.3 and 5.5 when the homogenization method and material model with rectangular holes were used. It seems that the high nonlinearity of the homogenization curves at regions near to the limit points is the reason. (See Figure 3.20)

Effect of the penalty exponent:

As noted earlier, the penalty exponent μ in (3.20) has a very important role in the nature of layout obtained. Relationship (3.20) can be written as

$$\mathbf{D}^H = V^\mu \mathbf{D}^0 \tag{5.1}$$

where $\mathbf{D}^H$ is the artificial constitutive matrix for the plane stress problem, V is the volume (area) of material and $\mathbf{D}^0$ is the isotropic elasticity matrix. The graph of V^{μ} (relative stiffness) versus volume, for different values of the penalty exponent μ, is illustrated in Figure 5.9. It is observed that by increasing the value of μ the curvature of the graph increases.

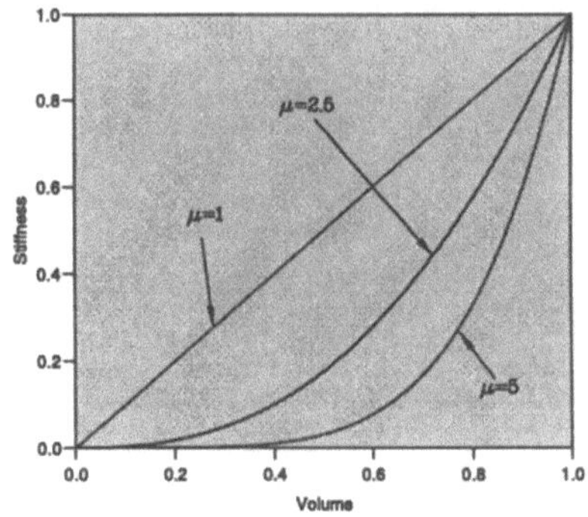

Figure 5.9 *A graph of V^{μ} (relative stiffness) against V for the the artificial material model.*

In the topological layout optimization problem the objective is to maximize the stiffness (or minimize compliance) for a given amount of material (the volume constraint). In Figure 5.9 it is seen that by increasing the value of μ, for the middle values of volume (i.e. middle values for a and b) smaller and smaller stiffnesses are obtained. Therefore, these middle values of a and b are suppressed in the optimization process and hence clearer black and white topological images are obtained.

In several papers by Rozvany *et al* [3-5] this fact is explained differently by introducing the concept of *solid isotropic microstructure with penalty* (SIMP) for intermediate volumes. He starts with an isotropic microstructure where the stiffness and volume are linearly dependent, then adds an artificial function which he calls the *cost of manufacturing of holes* to penalize the porous regions. In his terminology the volume of material is considered as a 'cost' to be added to the cost of manufacturing the holes.

To visualize the effect of the penalty exponent on the optimum layout, the problem of Figure 5.7 is now solved with different values of μ varying from 1.2 to 4.0. Other parameters are exactly as before. The layouts obtained after 130 iterations are depicted in Figure 5.10 (for $\mu = 4.0$ see Figure 5.7). It is observed that as the value of μ decreases more grey shaded areas appear. The iteration history is shown in Figure 5.11. (For $\mu = 4.0$ see Figure 5.8). It is observed that by decreasing the penalty exponent μ the range of variations of the objective function during the optimization process decreases.

As noticed before, the optimum values obtained from using different penalty

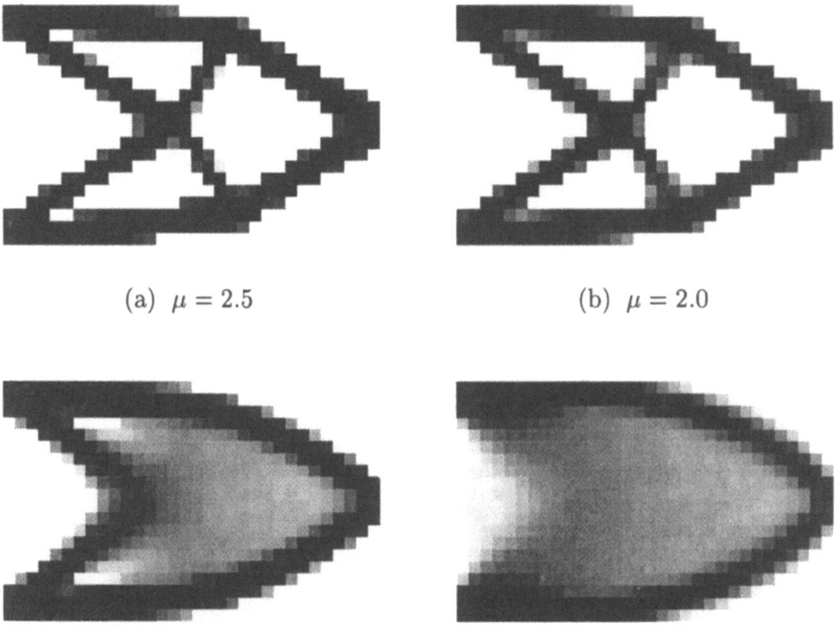

(a) $\mu = 2.5$ (b) $\mu = 2.0$

(c) $\mu = 1.5$ (d) $\mu = 1.2$

Figure 5.10 *Comparison of layouts obtained by using artificial material model with different penalty exponents.*

μ	1.0	1.2	1.5	2.0	2.5	4.0
Initial artificial strain energy	277.3	331.5	343.0	681.7	1073.4	4219.1
Initial real strain energy	1594.2	1594.2	1594.2	1594.2	1594.2	1594.2
Final artificial strain energy	198.9	220.3	245.5	251.7	254.0	255.9
Final real strain energy	256.0	245.6	238.9	240.4	241.2	243.3

Table 5.1 *Effect of the penalty exponent on the strain energy of the optimum layout when artificial material is used.*

exponents are not directly comparable. In Table 5.1 the initial and final (after

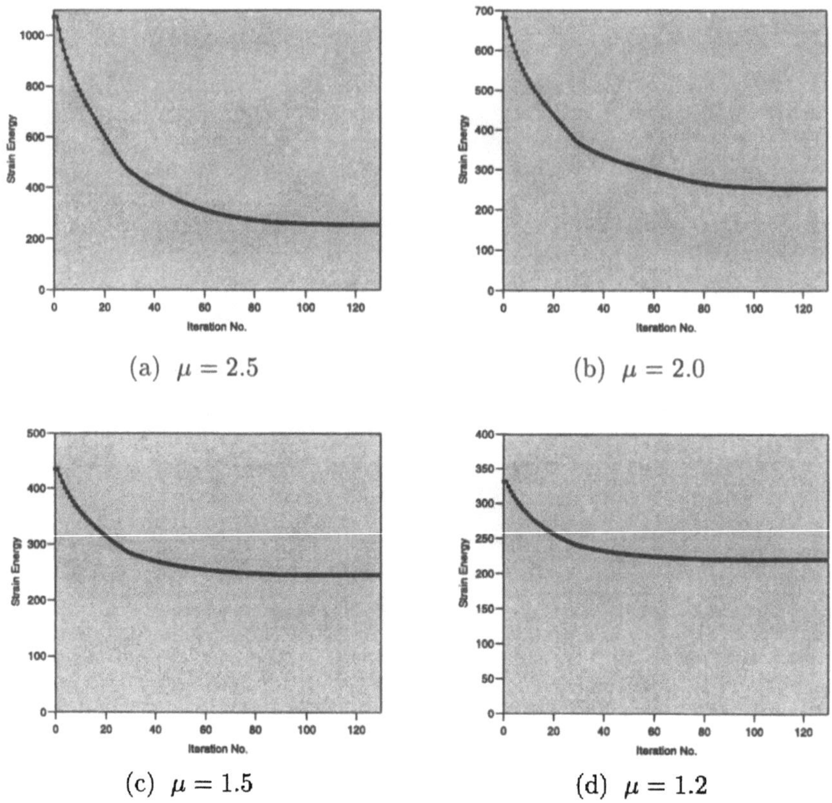

(a) $\mu = 2.5$ (b) $\mu = 2.0$

(c) $\mu = 1.5$ (d) $\mu = 1.2$

Figure 5.11 *Iteration history for the artificial material model with different penalty exponents.*

130 iterations) artificial and 'real' strain energies are compared. Note that the 'real' values are obtained using the homogenization method and the rectangular holes model with a bi-cubic least squares polynomial. From experiences with the artificial material with different penalty exponents, it is concluded that while $\mu \approx 1.5$ yields the smallest value for the strain energy, to obtain a clear image (i.e. solid-void layout) μ should be chosen between 3 and 4. The minimum strain energy value obtained in Table 5.1 is about 5.6% higher than the value obtained by homogenization.

5.2.3 Rank-2 material model

The rank-2 material model with the finite element mesh and other parameters exactly as for the cases considered earlier was used to solve the SCB problem. The layout obtained is shown in Figure 5.12. It seems that because this material has a very small value for shear strength a bracing-like pattern appears.

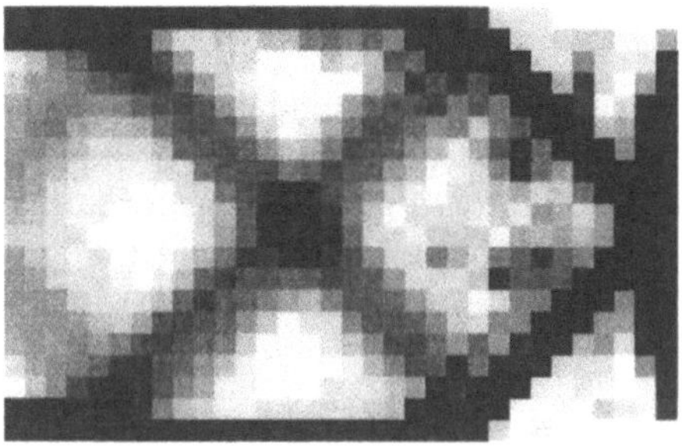

Figure 5.12 *Optimum layout obtained by using rank-2 material model.*

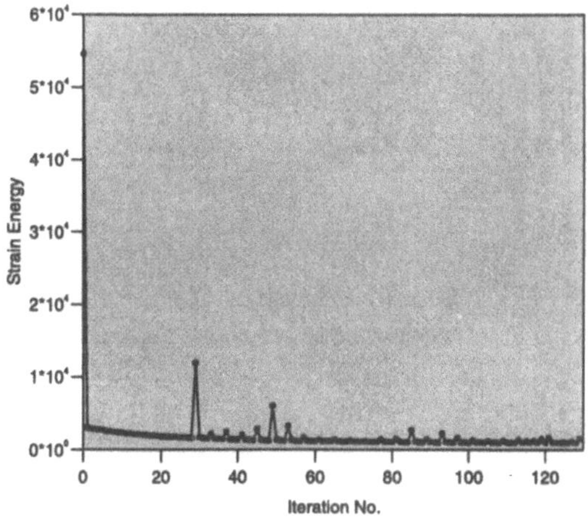

Figure 5.13 *Iteration history of SCB problem using rank-2 material model.*

It should be noted that to prevent singularity it is necessary to consider a very small value for the shear coefficient D_{66}. The adopted value is 0.00014. For a different loading case this structure might be unstable.

The iteration history is illustrated in Figure 5.13. It should be noted that the optimum value obtained for the strain energy using this material model can not be directly compared with other material models, because in contrast to the other material models this material does not have shear strength. Also note that by alignment of the material with the maximum principal direction a very large drop in the value of the strain energy has occurred. In fact, the orientation variable has a very important role which will be studied later.

5.3 Effect of the resizing scheme

In Section 4.5.2 a resizing scheme was explained. This scheme is similar to the one used by Suzuki *et al* [2] and Bendsøe *et al* [1]. In Section 4.5.3 a modified resizing scheme was introduced to obtain a monotonic algorithm. Here, the effect of this scheme on the obtained layout and the speed of convergence is studied.

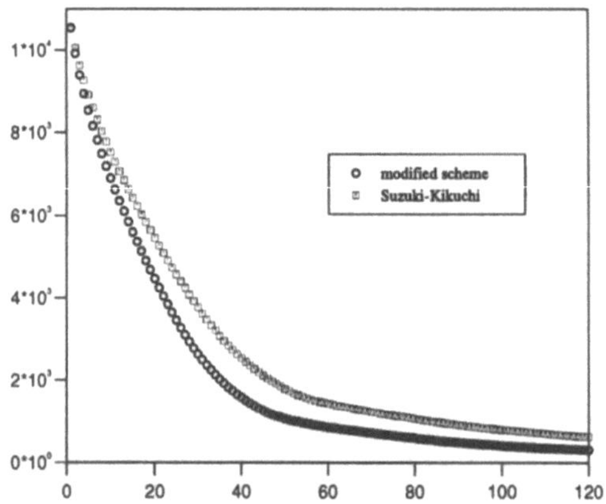

Figure 5.14 *Iteration history of SCB problem using two different resizing schemes.*

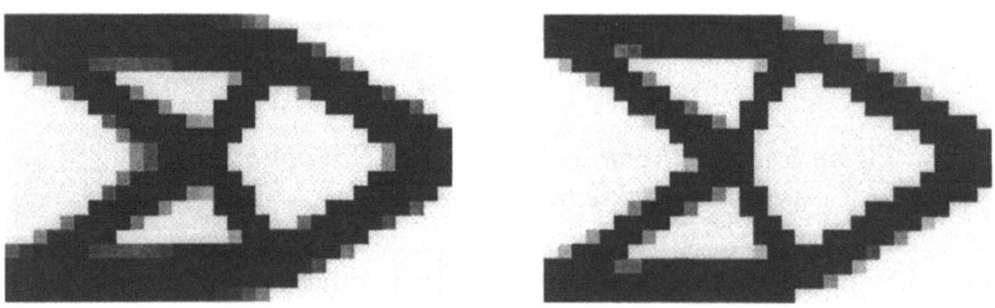

(a) (b)

Figure 5.15 *Layout obtained after 120 iterations using artificial material model and two different schemes: (a) Suzuki-Kikuchi's scheme and (b) modified scheme.*

The artificial material model is used in this study and the SCB problem is again considered. The move limit ζ and the volume tolerance δ are taken as 0.005 and 0.001, respectively. The penalty exponent μ is chosen to be equal to 4.0. When the resizing scheme of (4.73) and (4.75) is used, after 120 iterations the strain energy is 618.0, whereas using the modified scheme of (4.81) and (4.82) gives a value of 311.6. The iteration histories of these two experiments are shown in Figure 5.14. It is observed that the modified scheme has the faster convergence. The layouts obtained after 120 iterations are shown in Figure 5.15. In this chapter, unless otherwise stated, the modified resizing scheme is used.

5.4 Effect of the orientation variable

As noted earlier, when the homogenized material model (either rank-2 or cells with rectangular holes) are used, by correction of the orientation variable in the very first iteration a dramatic improvement in the objective function is observed. See, for example, Figures 5.4 and 5.13. In early works on structural topology optimization by Bendsøe and Kikuchi the influence of the rotation of the material was noticed [6].

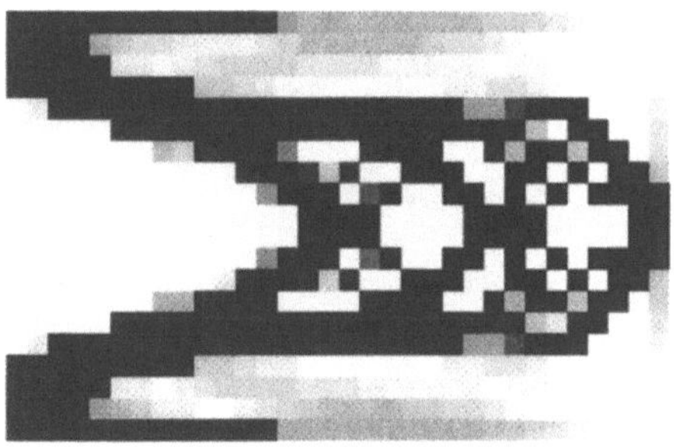

Figure 5.16 ***Layout obtained by using bi-cubic material model without considering the orientation variable.***

To study the effect of the orientation variable, the SCB problem was solved without consideration of the rotation angle θ^e as a design variable. When the bi-cubic version of the material model with rectangular voids was used the layout of Figure 5.16 was obtained. The strain energy after 130 iterations is 271.9, whereas when the orientation variable was included in the design variables the value obtained is 226.3.

As explained earlier, the orientation variable does not have any influence on the optimal layout when the isotropic artificial material model is employed.

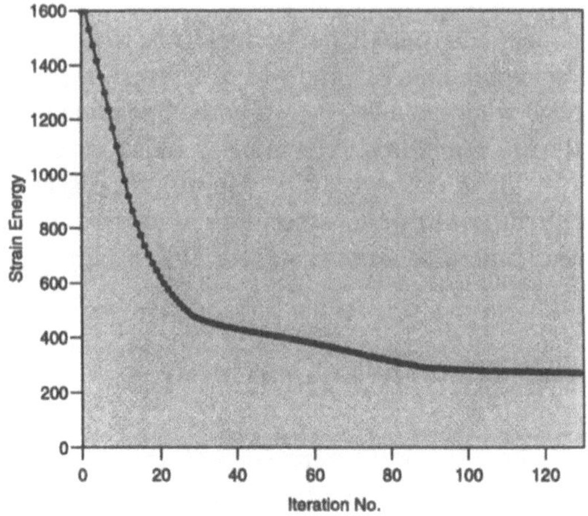

Figure 5.17 *Iteration history of SCB problem using bi-cubic material model without the orientation variable.*

Figure 5.18 *Layout obtained by using rank-2 material model without considering the orientation variable.*

The effect of the orientation variable when the rank-2 material model is used is crucial. In fact, without the orientation variable hardly any improvement in the objective function occurs. Figures 5.18 and 5.19 show the layout obtained and the iteration history. All the parameters are exactly the same as those adopted for the problem of Figure 5.12.

It should be noted that when the rank-2 material is used, a small value for the shear modulus has to be considered, otherwise the stiffness matrix will be singular. This small value for the shear modulus (D_{66} in (3.2)) is responsible

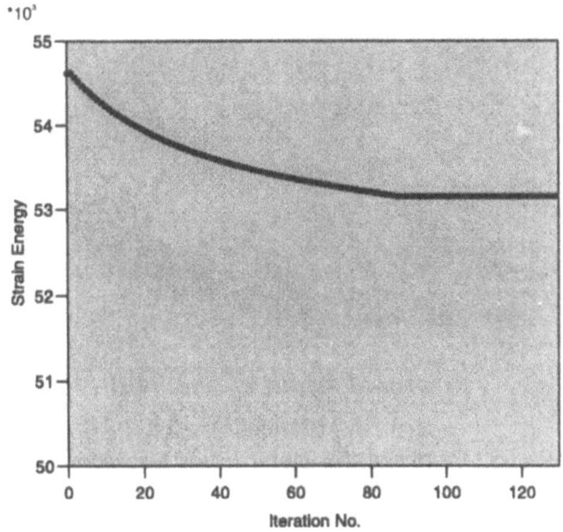

Figure 5.19 *Iteration history of SCB problem using rank-2 material model without the orientation variable.*

for the alignment of material in the principal stress direction. Otherwise, the orientation variable will always be equal to zero or $\pi/2$. Also, one should note that although the shear modulus is very small, the shear deformations of the rank-2 material are relatively large. As a general point, usually a truss like layout is obtained using the rank-2 material model.

5.5 Effect of the finite element discretization

To study the effect of the finite element discretization on the optimal layout and the problem of mesh dependency, the examples of Section 5.2 are now solved with a finer mesh. While the geometry and parameters and the type of elements of the SCB problem are kept as before, only the mesh resolution is increased from $32 \times 20 = 640$ eight noded elements to $64 \times 40 = 2560$ equal eight noded elements.

When the bi-cubic version of the material model with rectangular holes was used, after 130 iterations the layout of Figure 5.20 was obtained. The initial strain energy value was 1628.5 and after 130 iteration reduced to 225.7. The iteration history up to 200 iterations is depicted in Figure 5.21. At 200 iterations the strain energy value was 221.8 and after that no improvement occurred.

In the next example the bi-quartic version of the material model with rectangular holes was used. The starting value of the strain energy in this case was 1726.7 and after 130 iterations it reduced to 227.2. By continuing the process further hardly any improvement in the objective function was observed. The

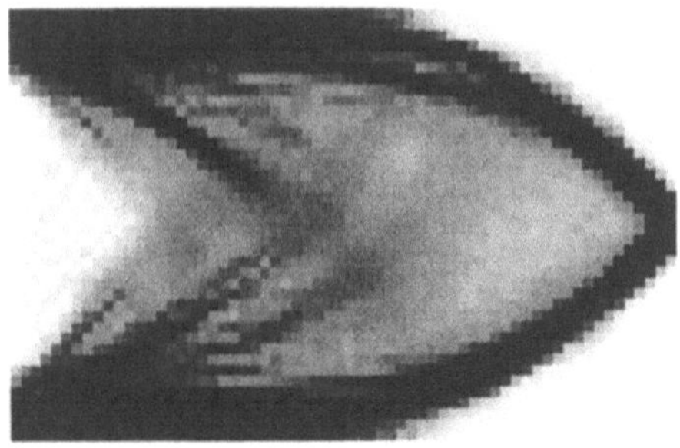

Figure 5.20 *Optimal layout of the SCB problem using a fine mesh and the material model with rectangular holes and bi-cubic polynomial.*

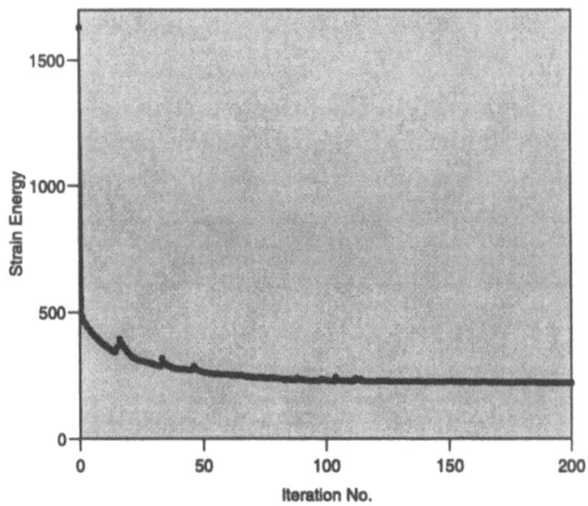

Figure 5.21 *Iteration history of SCB problem using a fine mesh and the material model with rectangular holes and bi-cubic least squares polynomial.*

optimal layout and the iteration history up to 200 iterations are illustrated in Figures 5.22 and 5.23 respectively.

To study the effect of the finite element mesh on the topology when the artificial material model is used, the SCB problem was solved with the above-mentioned fine mesh. A penalty exponent of $\mu = 4.0$ was adopted as in Section 5.2.2. The

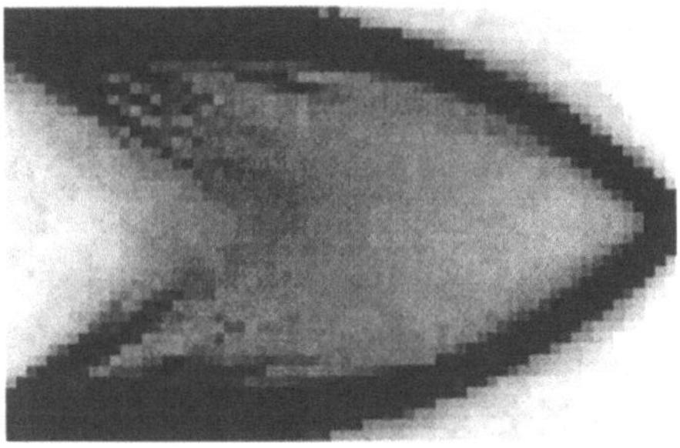

Figure 5.22 *Optimal layout of the SCB problem using a fine mesh and the material model with rectangular holes and bi-quartic polynomial.*

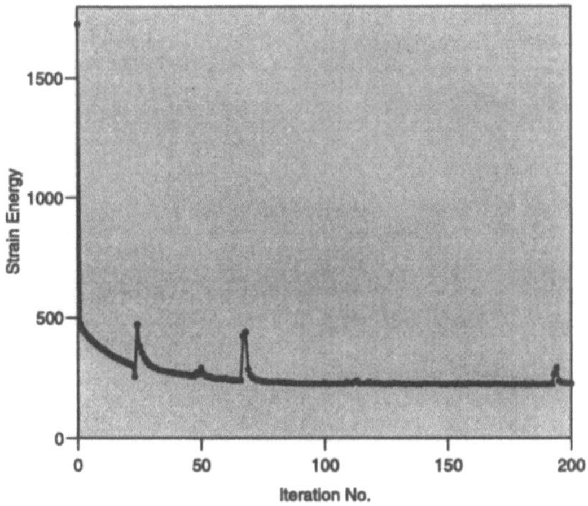

Figure 5.23 *Iteration history of SCB problem using a fine mesh and the material model with rectangular holes and bi-quartic polynomial.*

resulting topology is illustrated in Figure 5.24. It is observed that the layout obtained by using the fine mesh has two extra members. The results by various researchers (e.g. [1, 7, 8]) suggest that there is greater mesh dependency when the artificial material model is used.

The initial artificial strain energy was 4306.9 and after 130 iterations it reduced to 241.1 which is less than the value of 255.9 obtained using the coarse mesh. The graph of the objective function versus the number of iterations is

Figure 5.24 *Optimal layout of the SCB problem using a fine mesh and the artificial material model with* $\mu = 4.0$.

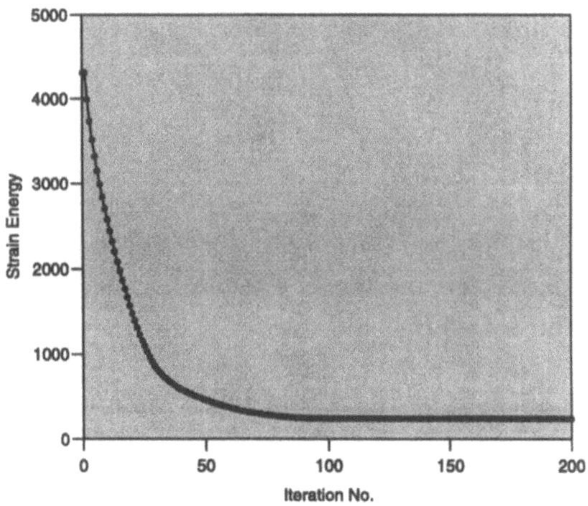

Figure 5.25 *Iteration history of SCB problem using a fine mesh and the artificial material model with* $\mu = 4.0$.

illustrated in Figure 5.25. As usual the convergence curve obtained is very smooth for the isotropic artificial material model.

To study the sensitivity of the rank-2 material on the finite element mesh, the SCB problem was solved using the above fine mesh. The resulting layout and the iteration history up to 200 iterations are depicted in Figures 5.26 and 5.27 respectively.

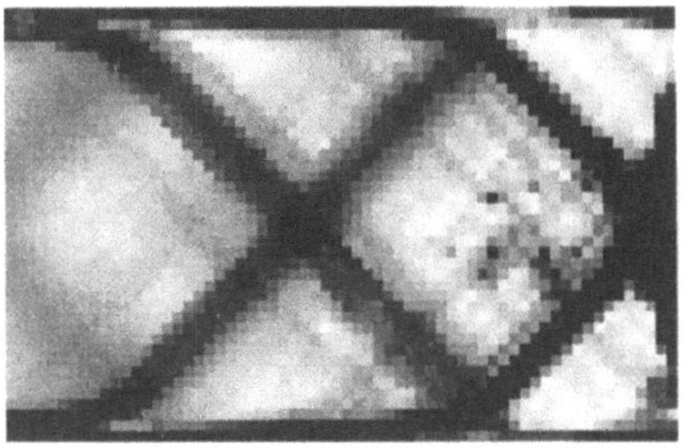

Figure 5.26 *Optimal layout of the SCB problem using a fine mesh and the rank-2 material model.*

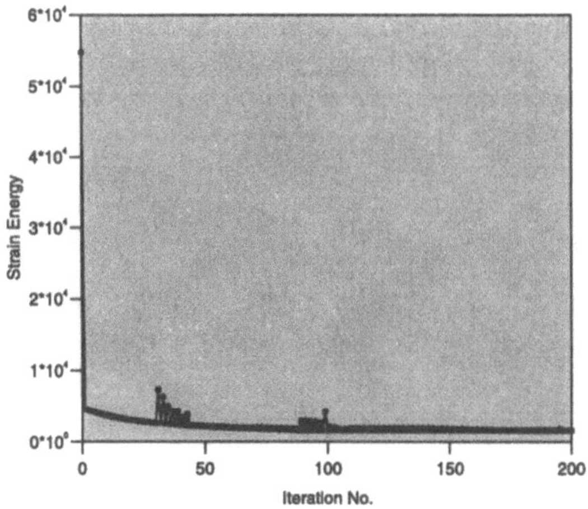

Figure 5.27 *Iteration history of SCB problem using a fine mesh and the rank-2 material model.*

5.5.1 Continuation method

As we noticed in the previous section, the resulting topology is dependent on the finite element mesh especially when the artificial material model used. See Figures 5.7 and 5.24. To decrease the effect of the finite element mesh on the optimal layout, the *continuation* approach was suggested by Jog [9]. In this approach the penalty exponent during the optimization process is gradually increased as the iterations proceed.

In an attempt to illustrate the effect of the continuation method it was used

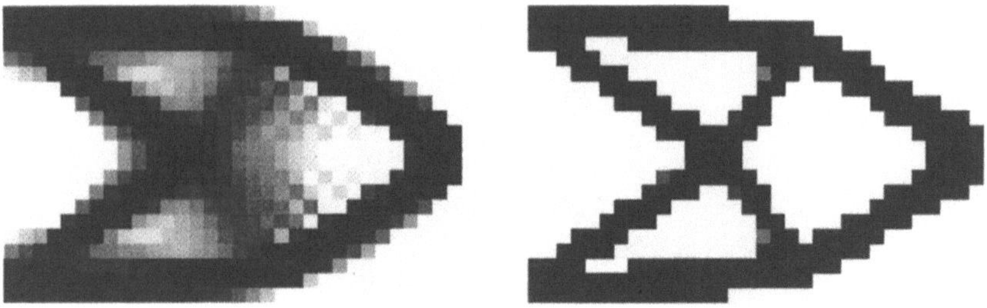

Figure 5.28 *Obtained layout for the SCB problem using the artificial material model and continuation approach after 50 (left) and 150 iterations (right). The employed mesh has* 32×20 *elements.*

to solve the SCB problem using the artificial material model. First the coarse mesh was used and in the second experiment the fine mesh was employed. The initial penalty exponent adopted was 1.5 and this was gradually increased to reach 4.0 after a hundred iterations. From there it was kept constant and equal to 4.0.

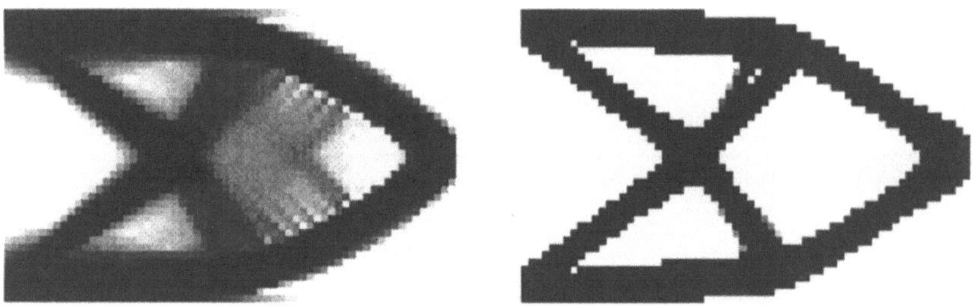

Figure 5.29 *Obtained layout for the SCB problem using the artificial material model and continuation approach after 50 (left) and 150 iterations (right). The employed mesh has* 64×40 *elements.*

The layouts obtained after 50 and 150 iterations using the coarser mesh (32×20 elements) are illustrated in Figure 5.28. It is seen that the topology obtained is similar to the one of Figure 5.7 where a constant penalty exponent was used. It should be noted that since the elasticity matrix of the artificial material is variable, the graph of the strain energy versus the iteration number in this approach is not very meaningful.

In Figure 5.29 the above experiment is repeated with the finer mesh. It seems that, in contrast with the layout of Figure 5.24, the topology of the fine mesh is the same as when the coarse mesh is used. The issue of mesh dependency will be dealt with further in Chapter 8.

5.5.2 Unstructured mesh

The use of unstructured grid generation techniques facilitates the meshing of more complex domains. As an example of an unstructured mesh the SCB problem was solved with the mesh of Figure 5.30. The mesh is produced using the ADOPT software [10] and it contains 1192 elements, all of which are 8-noded quadrilaterals. The mesh produced is not symmetric. The small rectangle in the middle of the free edge is considered as a non-design domain.

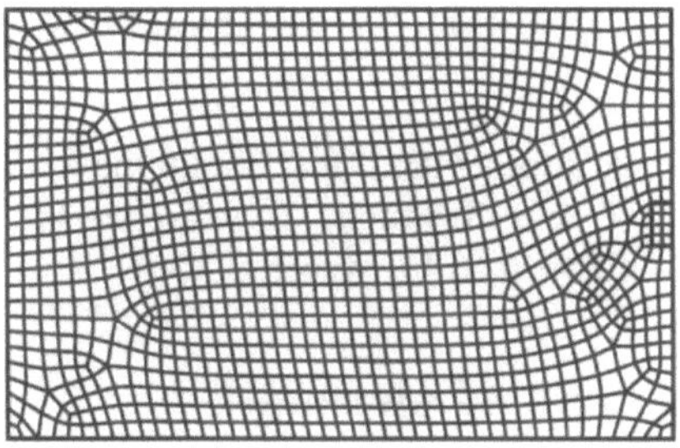

Figure 5.30 *Unstructured mesh for the SCB problem.*

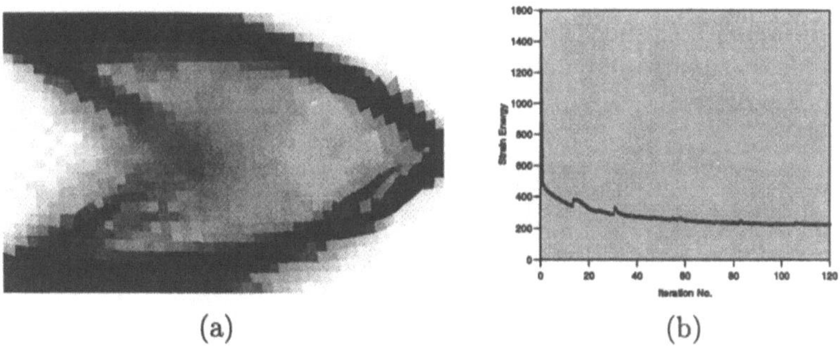

(a) (b)

Figure 5.31 *(a) Optimal layout for the SCB problem using the material model with rectangular holes and a bi-cubic polynomial. (b) The iteration history.*

Using the bi-cubic version of the material model with rectangular voids, the layout of the Figure 5.31(a) was obtained. The iteration history is shown in Figure 5.31(b). The resulting strain energy after 130 iterations is 225.5.

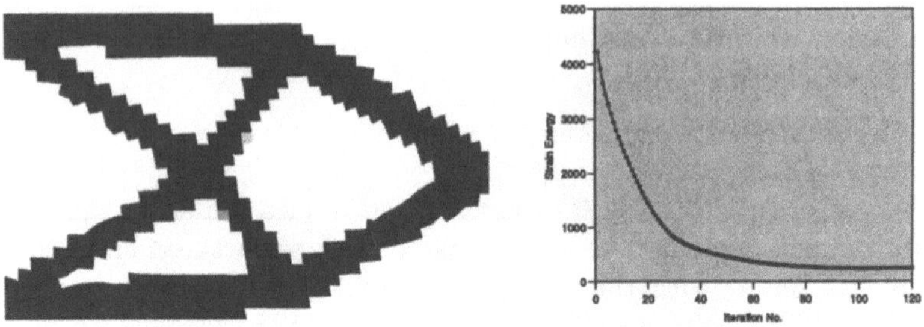

(a) (b)

Figure 5.32 *(a) Optimal layout for the SCB problem using the artificial material model with* $\mu = 4.0$. *(b) The iteration history.*

By repeating the above example with the artificial material model ($\mu = 4.0$) the topology of Figure 5.32(a) was obtained. To visualize the trend of convergence the iteration history is illustrated in Figure 5.32(b).

5.6 Effect of the type of elements

To demonstrate the effect of the use of a different type of element on the result of the topology optimization, 4-node linear elements are now considered. The result obtained by using the material model with rectangular holes and the bi-cubic least squares polynomial is depicted in Figure 5.33. A checkerboard pattern is obtained.

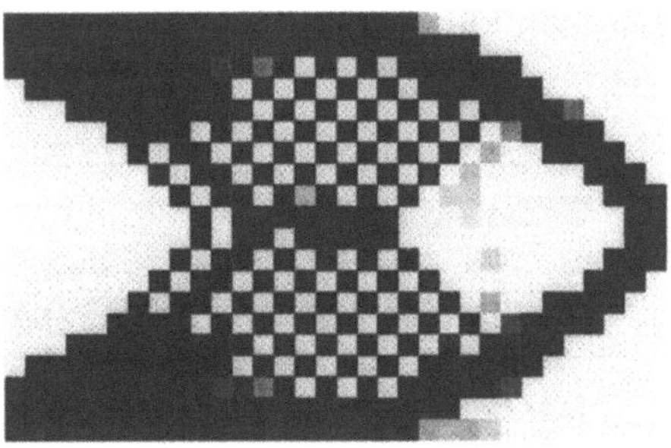

Figure 5.33 *Checkerboard pattern using 4-node elements.*

As another experiment, the 4-node elements and the artificial material model with $\mu = 5$ was tried. Again a checkerboard pattern can be recognized as shown in Figure 5.34.

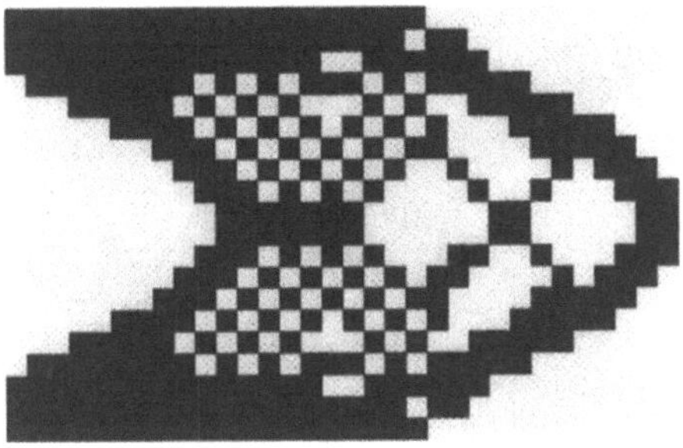

Figure 5.34 *Optimal layout using artificial material model and 4-node elements.*

The checkerboard problem has been addressed by Días *et al* [11] and Bendsøe *et al* [1].

5.7 Effect of the material volume

The value of the specified material volume in the design domain dramatically affects the optimal topology. To demonstrate this, the SCB problem was solved with two different volume fractions: 20% and 60%.

In the first study the artificial material model was adopted with a fine structured mesh of 64 × 40, 8-noded elements and the penalty exponent equal to 4.0. With the volume fraction equal to 20% and 60% the layouts of Figures 5.35 and 5.36 were obtained, respectively. In the latter case after 150 iterations the artificial and real (homogenized) strain energies are 165.6 and 163.1. Note that if the layout comprises only the solid material these two values should be equal.

Figure 5.35 *Optimal layout obtained using artificial material model and* $V_{\text{solid}}/V = 20\%$.

Figure 5.36 *Optimal layout using artificial material model and* $V_{\text{solid}}/V = 60\%$.

In the next study the material model with rectangular holes (bi-cubic version) was used. With the volume fraction $V_{\text{solid}}/V = 20\%$ the structure did not

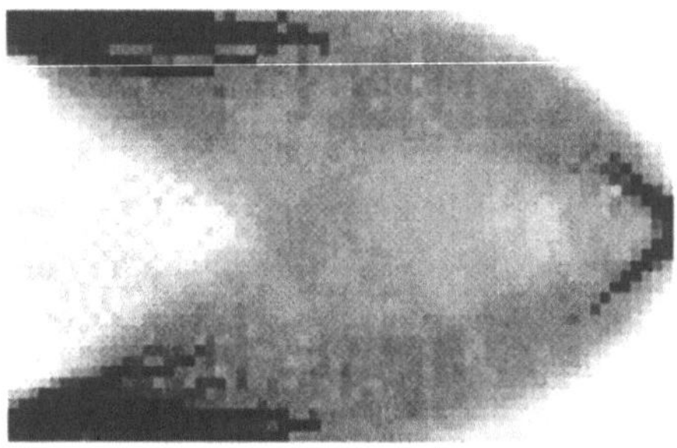

Figure 5.37 *Layout obtained using the material model with rectangular holes and* $V_{\text{solid}}/V = 20\%$.

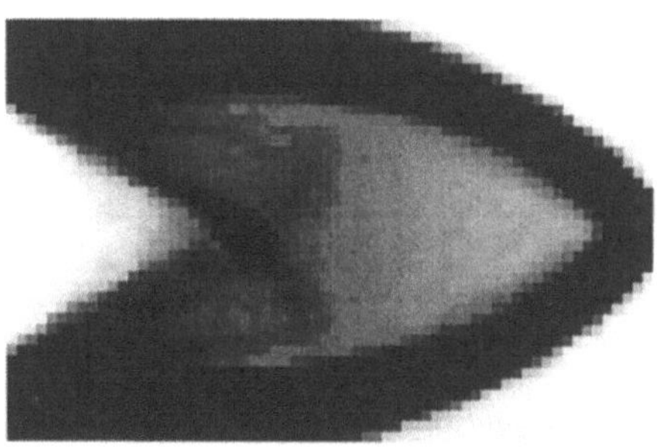

Figure 5.38 *Optimal layout using the material model with rectangular holes and* $V_{\text{solid}}/V = 60\%$.

develop properly (see Figure 5.37). When a volume fraction equal to 60% was used, the layout of Figure 5.38 was obtained. The strain energy value after 150 iterations was 156.9.

5.8 Effect of the resizing parameters

Amongst the parameters of the resizing scheme the damping factor η, used in (4.73) and (4.75), was discussed in Section 4.5.2. As we noticed this parameter does not exist in the suggested modified resizing scheme of (4.81) and (4.82). The acceptable tolerance for the active volume constraint δ and the move size ζ are studied here. See Section 4.5.2.

By studying the resizing schemes of Sections 4.5.2 and 4.5.3, it is can be easily realized that by choosing a larger value for the move size ζ the speed of convergence (i.e. the movement towards the optimum point in the feasible design space) increases. However, by increasing the move size, firstly, the chance of targetting the global minimum decreases and secondly, it is necessary to relax the accepted value for the tolerance of the equality volume constraint. Consequently, this reduces the accuracy. It should be noted that without paying attention to the dependency of the move size ζ and the volume tolerance δ one might encounter an unstable algorithm.

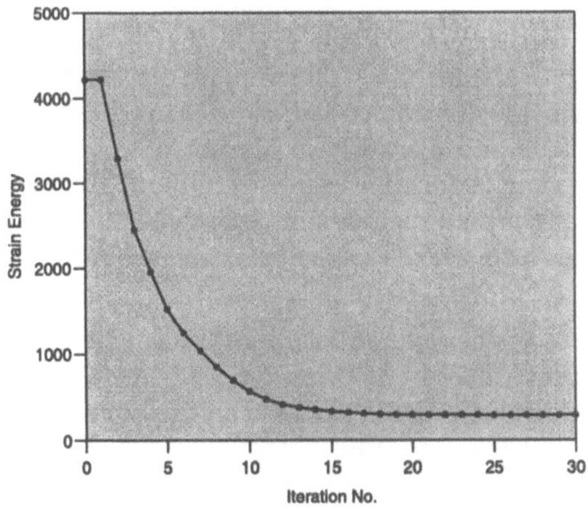

Figure 5.39 *Iteration history for the artificial material model with* $\mu = 4.0, \zeta = 0.05$ *and* $\delta = 0.1$.

To demonstrate the influence of the resizing scheme, the SCB problem is again considered. Because of its simplicity the isotropic artificial material model is employed. The penalty exponent is taken as $\mu = 4$. In Section 5.2.2 this problem was solved with the move size $\zeta = 0.008$ and the volume constraint $\delta = 0.001$. See Figures 5.7 and 5.8. In the first study the above problem was solved using $\zeta = 0.05$. Since this move size is relatively high, the allowed volume tolerance is increased to be $\delta = 0.1$. The iteration history is shown in Figure 5.39.

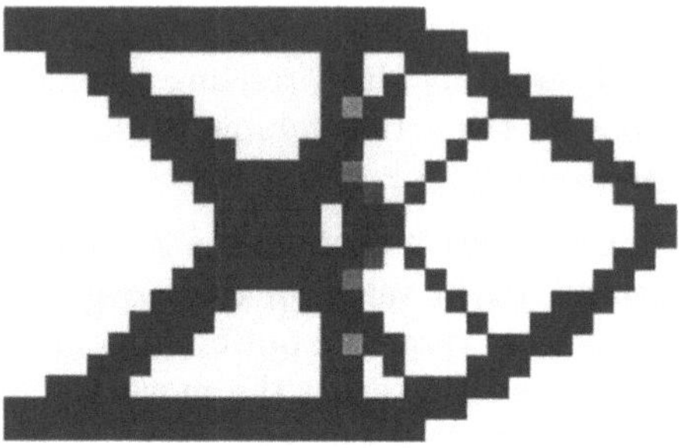

Figure 5.40 *Layout obtained using the artificial material model with* $\mu = 4.0, \zeta = 0.05$ *and* $\delta = 0.1$.

In this experiment in the first 20 iterations the strain energy value dropped from 4219.1 to 294.7 and after 25th iteration no further improvement occurred. When compared to the result of the experiment in Section 5.2.2 where the resizing values $\zeta = 0.008$ and $\delta = 0.001$ were used, we notice that the artificial strain energy after 130 iterations had reduced to 255.8, whereas using the parameters $\zeta = 0.05$ and $\delta = 0.1$ we are not able to get (artificial) strain energy values less than 292.3. However, it is noted that the convergence speed is much higher than for the experiment in Section 5.2.2. (Compare the iteration numbers in Figures 5.8 and 5.39). The resulting layout is illustrated in Figure 5.40.

To study the effect of the volume tolerance δ, the above experiment was repeated with different values of δ. Using $\zeta = 0.05$ and $\delta = 0.001$ the algorithm in the first few iterations became unstable and fell into a loop. By increasing the value of δ to 0.01 the algorithm became unstable after 22 iterations. The artificial strain energy value in the 22nd iteration is 298.8. In the next study δ was increased to 0.05. This time the algorithm was stable and the resulting layout and the iteration history are very similar to those of Figures 5.40 and 5.39. The final artificial strain energy value was 291.4 which is slightly smaller than for the case with $\delta = 0.1$. It seems that by decreasing the value of δ, apart from having greater accuracy for the constrained volume of material, the objective function also improves.

To increase the convergence rate, one may start with a large step size and decrease it whenever the algorithm becomes unstable.

5.9 Further examples

In this section to demonstrate the application of the method a few examples from structural engineering design are presented. In the first two examples the design of a bridge with four supports is considered. See Sections 5.9.1 and Section 5.9.2. In the next example the problem of optimum layout of a short cantilever beam or bracket with a centrally placed circular hole is solved. Finally, the problem of finding the optimal layout of a two-storey shear wall with openings is considered.

5.9.1 Bridge with support layout 1

This example is concerned with a bridge with uniformly distributed traffic loading. The dimensions of the design domain and the supports are illustrated in Figure 5.41. One row of elements at the level of the bridge deck is considered as a row of non-design elements. The following data are used: $E = 100\ 000$, $\nu = 0.3$ and $V_{\text{solid}}/V = 20\%$. Making use of symmetry only the left hand side of the problem is solved.

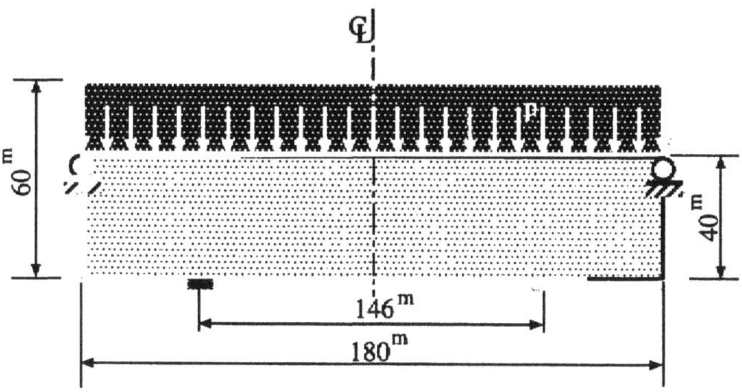

Figure 5.41 *Design domain of the bridge with support layout 1.*

Figure 5.42 *Optimum layout of the bridge problem using the material model with rectangular holes.*

Employing 8-node quadrilateral elements and the material model with rectangular voids and a complete bi-cubic least squares polynomial, a double arch bridge was obtained as illustrated in Figure 5.42.

When the rank-2 material is used, the image shown in Figure 5.43 emerges. Note that the problem is only solved for the symmetric left hand part of the bridge.

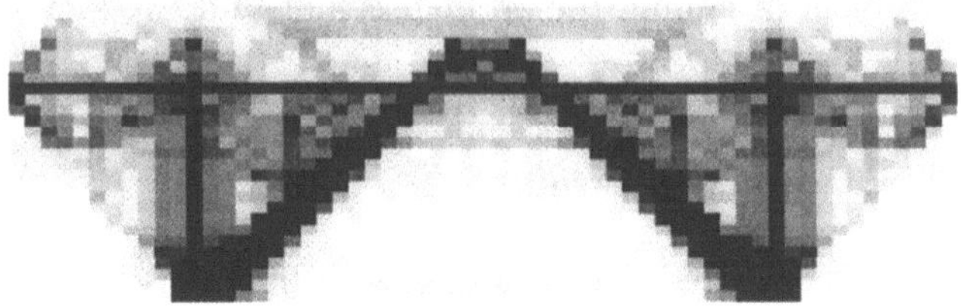

Figure 5.43 **Optimum layout of the bridge problem using the rank-2 material model.**

5.9.2 Bridge with support layout 2

Figure 5.44 shows a bridge with support layout 2 in which the supports were shifted to the bottom corners. The two rows of elements below the loading level were considered as non-design elements and again using symmetry only one half of the bridge was optimized.

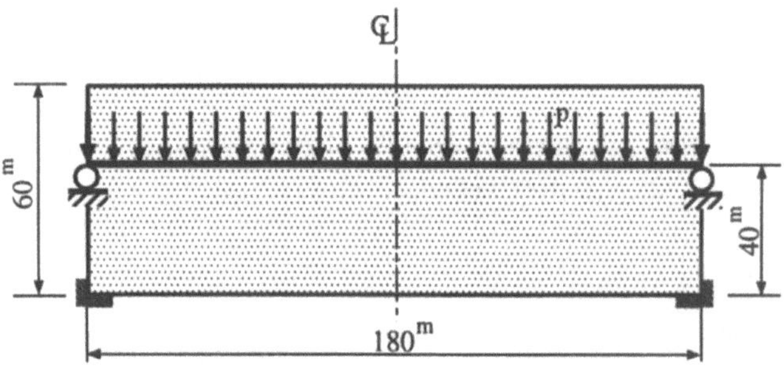

Figure 5.44 **Design domain of the bridge with support layout 2.**

In the first solution the material model comprising microcells with rectangular voids (bi-cubic) with a volume fraction of $V_{\text{solid}}/V = 40\%$ was examined. The optimum layout is depicted in Figure 5.45.

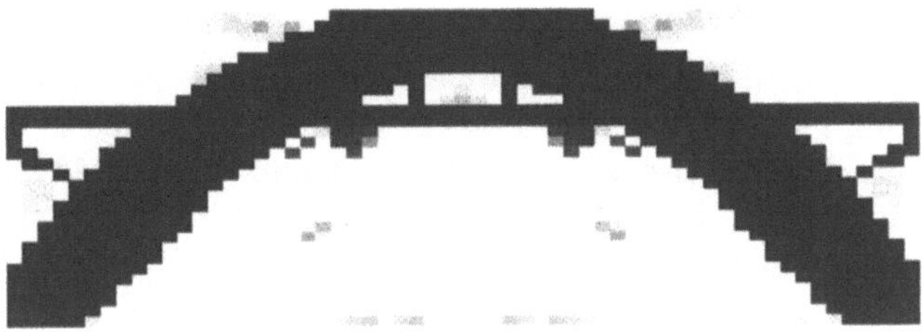

Figure 5.45 *Optimum layout of the bridge problem without middle supports using the material model with rectangular holes and* $V_{\text{solid}}/V = 40\%$.

Employing a rank-2 material with $V_{\text{solid}}/V = 15\%$ the structure in Figure 5.45 was obtained. Only one row of elements at the traffic level are considered here as non-design domain elements.

Figure 5.46 *Optimum layout of the bridge problem with rank-2 material model and* $V_{\text{solid}}/V = 15\%$.

The layout of Figure 5.47 was achieved using the artificial material model.

Figure 5.47 *Optimum layout of the bridge problem with the artificial material model.*

5.9.3 Bracket with a hole

For the bracket with a hole, the problem definition is illustrated in Figure 5.48. The Young's modulus and the Poisson's ratio are assumed to be 100000.0 and 0.3 respectively. The radius of the hole is 2.0 and it is located in the centre of the bracket. A point load of $P = 1000.0$ is applied in the top right corner. The left hand support is assumed to be fully clamped. $V_{\text{solid}}/V = 30\%$.

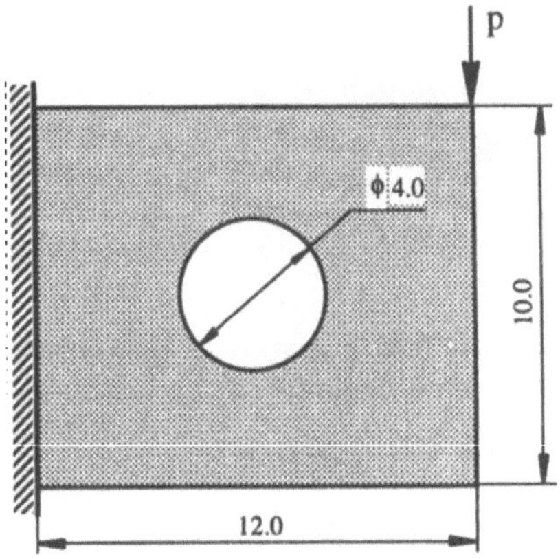

Figure 5.48 *Problem definition.*

In this problem the whole domain is considered as design domain. The domain is discretized by an unstructured mesh generator using the ADOPT software [10]. The finite element mesh is illustrated in Figure 5.49.

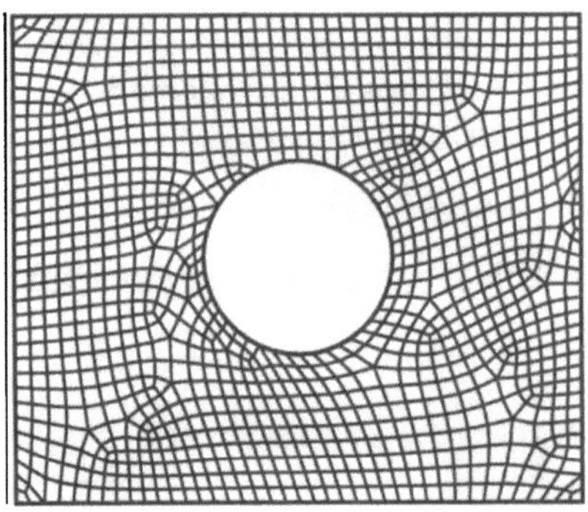

Figure 5.49 *Unstructured mesh for the bracket problem with a hole.*

Using the material model with rectangular holes and a complete bi-cubic least squares polynomial with the modified resizing scheme and the parameters $\zeta = 0.006$ and $\delta = 0.02$ the optimal layout of Figure 5.50 is obtained.

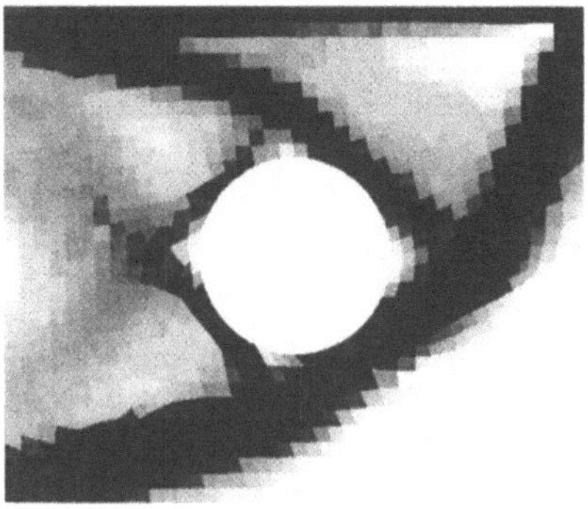

Figure 5.50 *Optimal layout for the bracket problem using the material model with rectangular voids.*

By using the isotropic artificial material model with the penalty exponent $\mu = 3.4$ the layout as illustrated in Figure 5.51 was obtained. The resizing parameters are as given above.

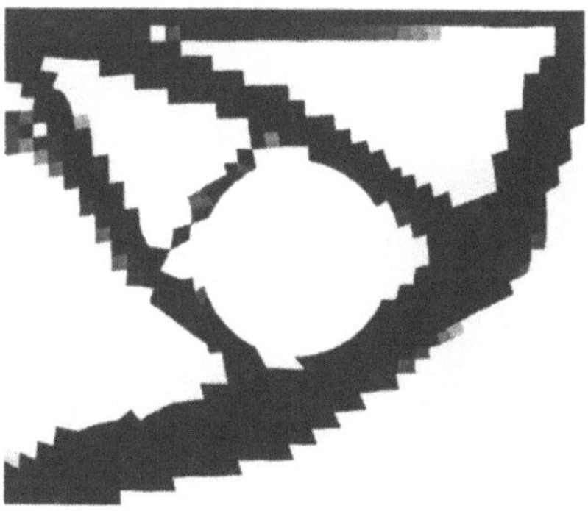

Figure 5.51 *Optimal layout for the bracket problem using the artificial material model.*

5.9.4 Shear wall with openings

The problem of finding the optimal layout of a shear wall when predefined openings exist is now considered. The problem definition is illustrated in Figure 5.52(a) where two horizontal point loads with intensity $P = 1000.0$ are applied. The resulting optimal layout for this loading (case 1) is shown in Figure 5.52(b).

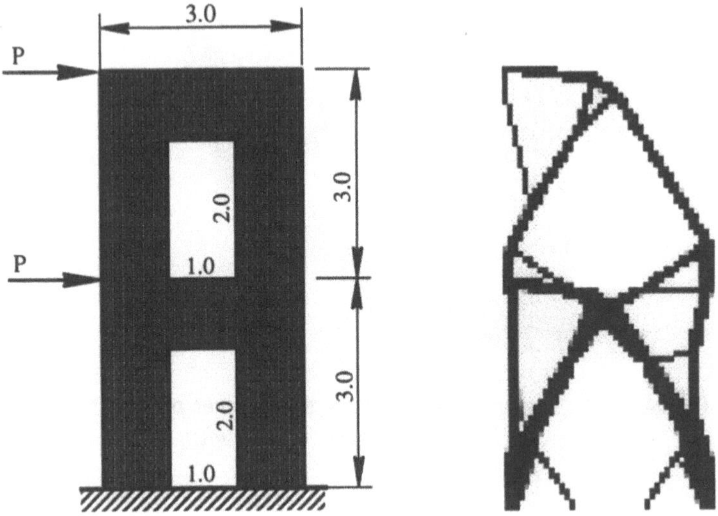

(a) (b)

Figure 5.52 *(a) Geometry and loading of the shear wall problem. (b) Optimal layout using the artificial material model.*

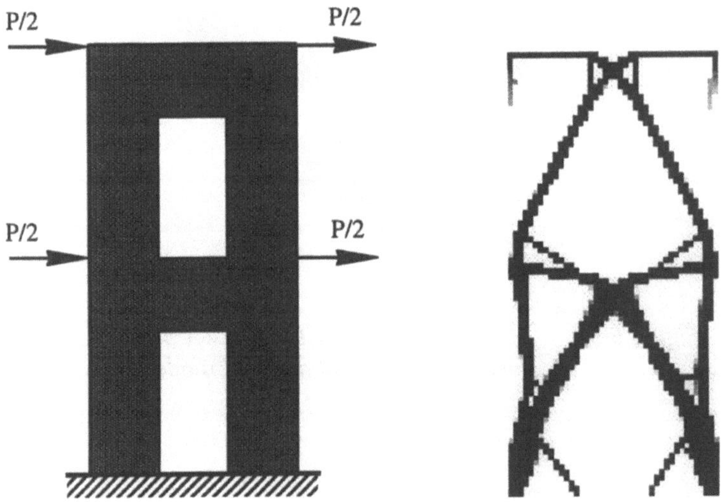

(a) (b)

Figure 5.53 *Shear wall problem. (a) Loading case 2. (b) Optimal layout.*

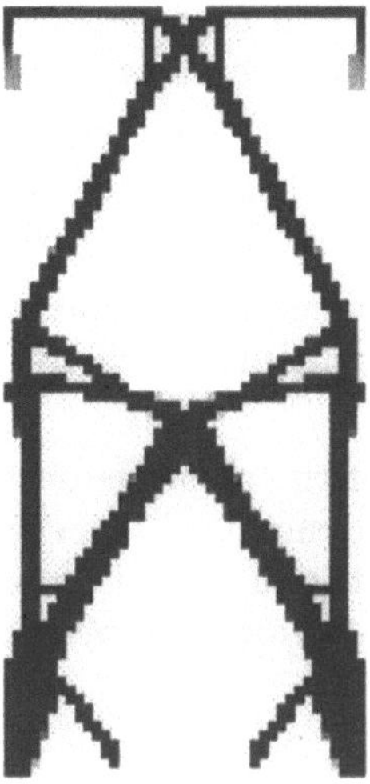

Figure 5.54 Optimal layout using the anti-symmetry condition.

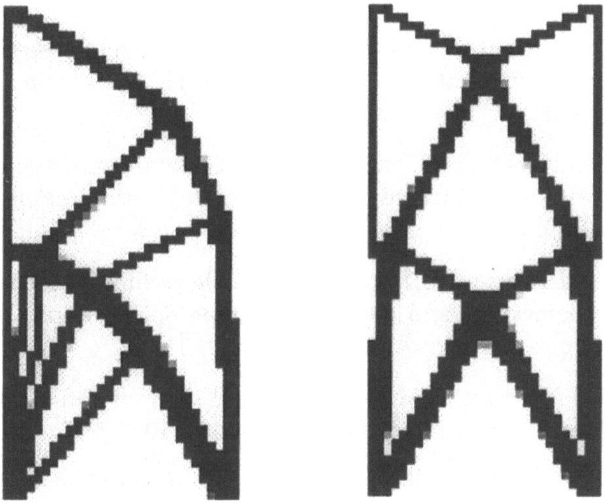

(a) (b)

Figure 5.55 Shear wall problem without openings. (a) Loading case 1. (b) Loading case 2.

The isotropic artificial material model with a penalty exponent of $\mu = 3.2$ and the modified resizing scheme with parameters $\zeta = 0.008$ and $\delta = 0.002$ were employed. The assumed volume fraction was taken to be $V_{\text{solid}}/V = 30\%$.

To study the effect of loading on the optimal topology this problem was solved for the loading case 2 shown in Figure 5.53(a). Using the same material model and parameters as before, the optimal layout of Figure 5.53(b) is obtained.

By making use of the anti-symmetry condition in load case 2, the problem was also solved for only half of the domain. The resulting layout is illustrated in Figure 5.54.

Finally, to study the effect of the openings on the optimal layout, this time a shear wall with the same dimensions as before without the openings was considered. Optimizing this structure for the loading cases 1 and 2 the layouts of Figures 5.55(a) and 5.55(b) are obtained, respectively. The problem is solved for the whole domain.

References

[1] Bendsøe M.P., Díaz A.R. and Kikuchi N., Topology and generalized layout optimization of elastic structures, in *Topology design of structures*, edited by Bendsøe M.P. and Mota Soares C. A., pp. 159–205. Kluwer Academic Publishers (1993)

[2] Suzuki K. and Kikuchi N., A homogenization method for shape and topology optimization, *Comp. Meth. Appl. Mech. Eng.*, **93**, 291–318 (1991)

[3] Rozvany G.I.N., Zhou M. and Birker T., Generalized shape optimization without homogenization, *Structural Optimization*, **4**, 250–252 (1992)

[4] Zhou M. and Rozvany G.I.N., The COC algorithm, part II: Topological, geometrical and generalized shape optimization, *Comp. Meth. Appl. Mech. Eng.*, **89**, 309–335 (1991)

[5] Rozvany G.I.N., Bendsøe M.P. and Kirsch U., Layout optimization of structures, *Appl. Mech. Rev.*, **48**(2), 41–119 (1995)

[6] Bendsøe M.P. and Kikuchi N., Generating optimal topologies in structural design using homogenization method, *Comp. Meth. Appl. Mech. Eng.*, **71**, 197–224 (1988)

[7] Bendsøe M.P., Optimal shape design as a material distribution problem, *Structural Optimization*, **1**, 193–202 (1989)

[8] Jog C.S., Haber R.B. and Bendsøe M.P., Variable topology-shape optimization with a constraint on the perimeter, in *20'th ASME Design Automatio Conference*, Minneapolis (1994)

[**9**] Jog C.S., *Variable-topology shape optimization of linear elastic structures*, Ph.D. thesis, Dept. of Theoretical and Applied Mechanics, Univ. of Illinois at Urbana-Champaign, USA (1993)

[**10**] Sienz J., *Integrated structural modelling, adaptive analysis and shape optimization*, C/ph/181/94, Dept. of Civil Engineering, University College of Swansea, Swansea, UK (1994)

[**11**] Díaz A. and Sigmund O., Checkerboard patterns in layout optimization, *Structural Optimization*, **10**, 40–45 (1995)

6

TOPOLOGICAL LAYOUT AND REINFORCEMENT OPTIMIZATION OF PLATE STRUCTURES

In this chapter the homogenization method for plane stress problems is extended to deal with the creation of optimal layout and reinforcement topologies of plate structures. In this formulation the problem is to determine the stiffest plate structure with a volume constraint for the reinforcement material. Introducing plate microcell models, a formulation for homogenization of plates based on the first order thick plate theory is presented. To solve the optimization problem the optimality criteria method is used. Using different material models some examples are provided.

6.1 Introduction

Plates are one of the most important structural systems used in engineering practice. The problem of optimum layout and reinforcement of plates has attracted the attention of researchers since the late 70's. The early results showed that the optimum reinforcement of Kirchhoff plates may include an infinite number of fine ribs in some areas [1-3]. Cheng and Olhoff [3] studied the plate optimization problem under uniform loading using the optimality criteria method and a finite difference scheme. Rozvany *et al* [4] realized that while unconstrained optimization, in general, leads to plates with smooth thickness variation, bounding the design variables results in designs with discrete ribs of various sizes. Bendsøe [5] used homogenization techniques in plates with ribs in one direction and applied a 'smeared-out' procedure to compute the homogenized properties of plates with ribs in two orthogonal directions.

Models based on Mindlin-Reissner plates have been studied by Bourgeat and Tapiero [6], Suzuki and Kikuchi [7], Tenek and Hagiwara [8, 9], Soto and Díaz [10] and Díaz *et al* [11].

To establish a well-posed and relaxed formulation, 'smeared-out' techniques and more specifically, the homogenization method may be employed. Suzuki and Kikuchi [7, 12] applied the homogenization method for plate and shell structures. They introduced an approach to compute homogenized plate stiffness properties by using the homogenized properties derived from two dimensional elasticity. Soto and Díaz [10] applied the homogenization techniques directly to the plate equation to obtain homogenization properties without the 'smeared-out' technique of Bendsøe or homogenized properties from two dimensional elasticity.

Recently, Díaz *et al* [11] have suggested a different method that uses optimal layered microstructures of finite rank. In this method the optimal properties of layered material is determined 'locally' and the amount of material allocated to build the local anisotropic microstructures is determined 'globally' within the optimization scheme. They have used the sequential quadratic programming (SQP) algorithm for the local problem and optimality criteria methods for the global problem.

In this chapter, we will present a method similar to Suzuki and Kikuchi [7] and Bendsøe *et al* [13] with some modification for the shear rigidity coefficients. For optimization problem an optimality criteria method is employed, where the objective function is the total potential energy to be maximized. A global volume constraint is considered for the allocated reinforcement. According to the developed algorithm, it is possible to design plates with external reinforcement (ribbed plates), internal reinforcement (honeycomb plates) and transversely perforated plates. As with the plane stress problems, different material models such as the microcells with rectangular holes, the rank-2 and the artificial materials can be used.

For the plate formulation Mindlin-Reissner thick plate theory is here adopted. For finite element analysis, a hierarchical, heterosis plate element is employed. It has already shown that this element performs well in thin as well as thick plate situations [14-16].

6.2 Selection of plate base cell model

From theoretical studies we know that in the problem of topological plate reinforcement or layout optimization, infinitesimal cavities or ribs need to be introduced. Using the homogenization method and introducing cellular microstructures assures the existence of solution and a relaxed and well-posed problem is obtained [17]. By employing the concept of assembling plies [7], the cell model of Figure 6.1 can be used [10].

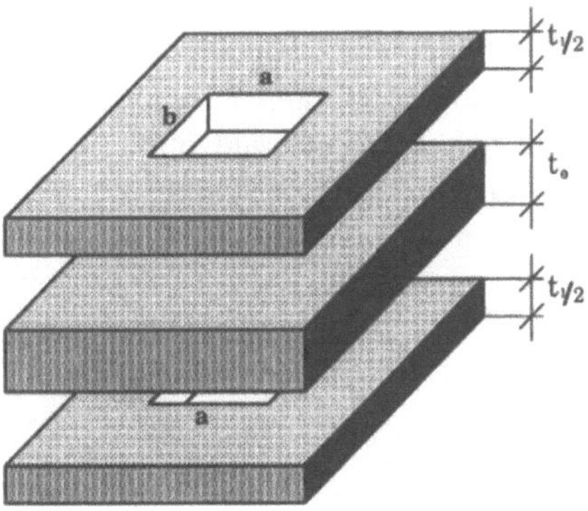

Figure 6.1 *Base cell of plate with ribbed reinforcement.*

In this model the main plate is the middle isotropic ply and the top and bottom layers comprise the reinforcement. For the reinforcement layers, to have a perforated microstructure, the material models introduced in Chapter 3 can be used.

If the material model with rectangular holes is used for the reinforcement, then as with the plane stress problem, the cell parameters are the sizes of the holes and the orientation of the hole. Hence, assuming that in each finite element the top and bottom layers consist of a unique material, the design variables of the optimization problem in each finite element are (a^e, b^e, θ^e). It should be noted that in this model it is assumed that t_0 the thickness of the main plate and t_1 the thickness of the reinforcement layers are constant while the sizes of the microscopic holes a and b, vary in the optimization process. Therefore, the size, location and orientation of 'ribs' (with a constant height) are determined by means of the topological reinforcement optimization algorithm.

It should be noted that for the sake of symmetry, the reinforcement layers are considered on the both sides of the plate (top and bottom). The allocated volume for the reinforcement material, in a continuum form, can be written as

$$V_s = t_1 \int_{\Omega_0} \big(1 - a(x,y)b(x,y)\big) d\Omega \tag{6.1}$$

where Ω_0 is the domain of the middle plane of the initial plate.

The reinforcement layer can be modelled using other material models such as rank layered materials or artificial materials. It is also interesting to note that the rank-1 material can be used here without the risk of inducing a singularity in the global stiffness matrix.

If the thickness of the middle ply t_0 is considered to be zero (this is equivalent to having only one cellular layer with thickness t_1), the topological optimization procedure will result in a grid-like perforated plate which is sometimes called the 'optimal plate layout'. This can also be achieved by considering three layers of cellular (homogenized) material.

It is possible to make another plate model by using the cellular homogenized material in the core and the isotropic solid plies as the face laminates. This model is illustrated in Figure 6.2 and it is sometimes called a 'honeycomb plate' [10].

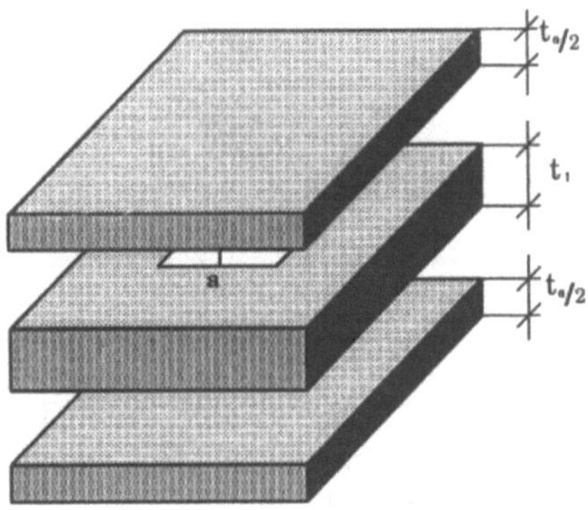

Figure 6.2 *Base cell of honeycomb plate model.*

The base cell models introduced in this section, can also be employed in shell problems [7, 13]. In this chapter, having in mind the possibility of future extension of the algorithm to shell problems, the formulation is presented in a more general form.

6.3 A brief review of Mindlin-Reissner plate theory

A plate occupies a three dimensional domain defined as:

$$\Omega = \left\{ (x, y, z) \in I\!R^3 | \ (x, y) \in \Omega_0 \subset I\!R^2 \ , \ z \in \left[-\frac{t}{2}, \frac{t}{2} \right] \right\}. \tag{6.2}$$

where t is the plate thickness and Ω_0 is the mid-plane about which the plate is symmetric. Although, no plate theory is completely consistent with the three-dimensional theory, the justification of the popular plate theories lie in their simplicity and usefulness in practical structural engineering applications.

Mindlin-Reissner plate theory allows for transverse shear deformation and unlike the classical Kirchhoff thin plate theory requires only C^0-continuity in finite element displacement formulations. The main assumptions are that:

1. Displacements are small compared to the plate thickness.
2. Stresses normal to the mid-surface are negligible ($\sigma_z = 0$).
3. Normals to the mid-surface before deformation remain straight but not necessarily normal to the mid-surface after deformation.

The displacement field resulting from the third assumption can be written as

$$\begin{aligned} u(x, y, z) &= u^0(x, y) + z\theta_x(x, y) \\ v(x, y, z) &= v^0(x, y) + z\theta_y(x, y) \\ w(x, y, z) &= w^0(x, y) \end{aligned} \tag{6.3}$$

where u, v and w are displacements in the x, y and z directions, respectively, and u^0, v^0 and w^0 are the associated mid-plane displacements. The variables θ_x and θ_y are the rotations in the xz and yz planes respectively due to bending. The sign convention for the displacements and generalized stress resultants of a typical Mindlin-Reissner plate under general loading is shown in Figure 6.3.

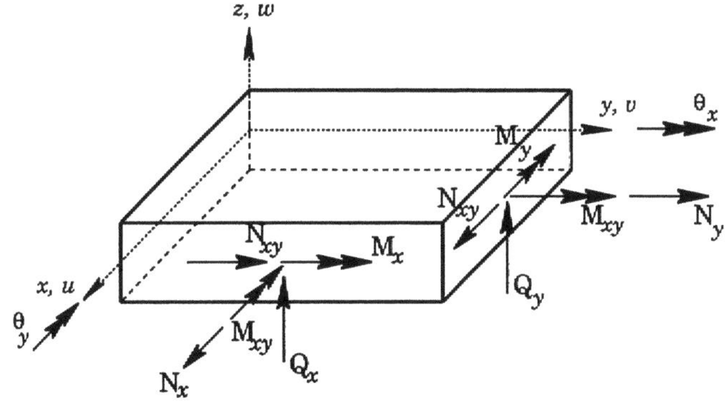

Figure 6.3 *A typical Mindlin-Reissner plate and sign convention for displacements and stress resultants.*

By making use of the assumption $\sigma_z = 0$, and since the stresses and strains in planes parallel to and perpendicular to the mid-plane are uncoupled, the elastic constitutive equations for an orthotropic material with axes of orthotropy in the x, y and z directions, at a typical point (x, y, z) may be written as

$$\begin{Bmatrix} \sigma_x \\ \sigma_y \\ \sigma_{yz} \\ \sigma_{xz} \\ \sigma_{xy} \end{Bmatrix} = \begin{bmatrix} C_{11} & C_{12} & 0 & 0 & C_{16} \\ C_{12} & C_{22} & 0 & 0 & C_{26} \\ 0 & 0 & C_{44} & C_{45} & 0 \\ 0 & 0 & C_{54} & C_{55} & 0 \\ C_{16} & C_{26} & 0 & 0 & C_{66} \end{bmatrix} \begin{Bmatrix} \epsilon_x \\ \epsilon_y \\ \gamma_{yz} \\ \gamma_{xz} \\ \gamma_{xy} \end{Bmatrix}. \tag{6.4}$$

It is convenient to separate the in-plane (plane stress problem) and transverse shear effects which leads to

$$\begin{Bmatrix} \sigma_x \\ \sigma_y \\ \sigma_{xy} \end{Bmatrix} = \begin{bmatrix} C_{11} & C_{12} & C_{16} \\ C_{12} & C_{22} & C_{26} \\ C_{16} & C_{26} & C_{66} \end{bmatrix} \begin{Bmatrix} \epsilon_x \\ \epsilon_y \\ \gamma_{xy} \end{Bmatrix} \tag{6.5a}$$

or

$$\boldsymbol{\sigma} = \mathbf{C}\boldsymbol{\epsilon}, \tag{6.5b}$$

and

$$\begin{Bmatrix} \sigma_{yz} \\ \sigma_{xz} \end{Bmatrix} = \begin{bmatrix} C_{44} & C_{45} \\ C_{54} & C_{55} \end{bmatrix} \begin{Bmatrix} \gamma_{yz} \\ \gamma_{xz} \end{Bmatrix}, \tag{6.6}$$

or

$$\boldsymbol{\sigma}_s = \mathbf{C}_s\boldsymbol{\gamma} \tag{6.7}$$

Using (6.3), strains can be expressed as

$$\begin{aligned} \epsilon_x &= \frac{\partial u^0}{\partial x} + z\frac{\partial \theta_x}{\partial x} = \epsilon_x^0 + z\chi_x, \\ \epsilon_y &= \frac{\partial v^0}{\partial y} + z\frac{\partial \theta_y}{\partial y} = \epsilon_y^0 + z\chi_y, \\ \gamma_{xy} &= \left(\frac{\partial u^0}{\partial y} + \frac{\partial v^0}{\partial x}\right) + z\left(\frac{\partial \theta_x}{\partial y} + \frac{\partial \theta_y}{\partial x}\right) = \gamma_{xy}^0 + z\chi_{xy} \end{aligned} \tag{6.8}$$

where χ_x and χ_y are the bending curvatures and χ_{xy} is the twisting curvature. Equation (6.8) may also be written as

$$\boldsymbol{\epsilon} = \boldsymbol{\epsilon}^0 + z\boldsymbol{\chi}. \tag{6.9}$$

where $\boldsymbol{\epsilon}^0 = [\partial u^0/\partial x, \partial v^0/\partial y, (\partial u^0/\partial y + \partial v^0/\partial x)]$ are the mid-plane strains induced by membrane forces and $\boldsymbol{\chi} = [\partial\theta_x/\partial x, \partial\theta_y/\partial y, (\partial\theta_x/\partial y + \partial\theta_y/\partial x)]$ are generalized bending strains which are measures of bending deformation. In (6.9) using the principal of superposition, we can deal with the bending and membrane effects separately. By doing so, the constitutive relationship can be written as

$$\boldsymbol{\sigma}^0 = \mathbf{C}\,\boldsymbol{\epsilon}^0 \tag{6.10}$$

for the membrane portion and

$$\sigma = z\,\mathbf{C}\,\chi \tag{6.11}$$

for the bending part.

Similarly, for the transverse shear strains we can write

$$\begin{aligned}\gamma_{yz} &= \frac{\partial v}{\partial z} + \frac{\partial w}{\partial y} = \theta_y + \frac{\partial w^0}{\partial y} ,\\ \gamma_{xz} &= \frac{\partial u}{\partial z} + \frac{\partial w^0}{\partial x} = \theta_x + \frac{\partial w^0}{\partial x} .\end{aligned} \tag{6.12}$$

The bending moments, membrane forces and shear forces as shown in Figure (6.3) are defined by the expressions

$$\mathbf{M} = \langle M_x, M_y, M_{xy}\rangle^T = \int_{-t/2}^{t/2} z\langle \sigma_x, \sigma_y, \sigma_{xy}\rangle^T dz, \tag{6.13}$$

$$\mathbf{N} = \langle N_x, N_y, N_{xy}\rangle^T = \int_{-t/2}^{t/2} \langle \sigma_x^0, \sigma_y^0, \sigma_{xy}^0\rangle^T dz, \tag{6.14}$$

$$\mathbf{Q} = \langle Q_x, Q_y\rangle^T = \kappa \int_{-t/2}^{t/2} \langle \sigma_{yz}, \sigma_{xz}\rangle^T dz \tag{6.15}$$

where κ is shear correction factor.

By substituting for the bending stresses in (6.13) from (6.11), the bending moments may be expressed as

$$\mathbf{M} = \int_{-t/2}^{t/2} z^2 \mathbf{C}\,\chi\, dz, \tag{6.16}$$

or

$$\begin{Bmatrix} M_x \\ M_y \\ M_{xy} \end{Bmatrix} = \begin{bmatrix} D_{11} & D_{12} & D_{16} \\ D_{12} & D_{22} & D_{26} \\ D_{16} & D_{26} & D_{66} \end{bmatrix} \begin{Bmatrix} \chi_x \\ \chi_y \\ \chi_{xy} \end{Bmatrix} \tag{6.17}$$

or

$$\mathbf{M} = \mathbf{D}_f\, \chi, \tag{6.18}$$

where $\mathbf{D}_f$ is the flexural rigidity matrix of plate which has the form

$$\mathbf{D}_f = \int_{-t/2}^{t/2} z^2\, \mathbf{C}\, dz. \tag{6.19}$$

Similarly the membrane forces may be written as

$$\mathbf{N} = \mathbf{D}_m\, \epsilon^0 , \tag{6.20}$$

where $\mathbf{D}_m$ is the membrane (extensional) stiffness matrix. Furthermore, if we substitute for the membrane stresses in (6.14) from (6.10) by comparing with (6.20) it is concluded that

$$\mathbf{D}_m = \int_{-t/2}^{t/2} \mathbf{C}\, dz\,. \tag{6.21}$$

Following this procedure for the transverse shear forces, we can write

$$\mathbf{Q} = \mathbf{D}_s \gamma \tag{6.22}$$

where the matrix of shear rigidities $\mathbf{D}_s$ is obtained as

$$\mathbf{D}_s = \kappa \int_{-t/2}^{t/2} \mathbf{C}_s\, dz. \tag{6.23}$$

Note that $\gamma = \left[\gamma_{yz}, \gamma_{xz}\right]^T$.

For an isotropic plate with thickness t, Young's modulus E and Poisson's ratio ν the constitutive relations become

$$\mathbf{D}_f = \frac{Et^3}{12(1-\nu^2)} \begin{bmatrix} 1 & \nu & 0 \\ \nu & 1 & 0 \\ 0 & 0 & (1-\nu)/2 \end{bmatrix}, \tag{6.24}$$

$$\mathbf{D}_m = \frac{Et}{1-\nu^2} \begin{bmatrix} 1 & \nu & 0 \\ \nu & 1 & 0 \\ 0 & 0 & (1-\nu)/2 \end{bmatrix}, \tag{6.25}$$

and

$$\mathbf{D}_s = \frac{\kappa E t}{2(1+\nu)} \begin{bmatrix} 1 & 0 \\ 0 & 1 \end{bmatrix}. \tag{6.26}$$

The shear correction factor κ is usually taken between $2/3$ and $5/6$ to allow for warping of the cross-section.

The constitutive relations for the plate problem in a general and compact form can be summarized as

$$\begin{Bmatrix} \mathbf{N} \\ \mathbf{M} \\ \mathbf{Q} \end{Bmatrix} = \begin{bmatrix} \mathbf{D}_m & 0 & 0 \\ 0 & \mathbf{D}_f & 0 \\ 0 & 0 & \mathbf{D}_s \end{bmatrix} \begin{Bmatrix} \epsilon^0 \\ \chi \\ \gamma \end{Bmatrix}. \tag{6.27}$$

In the next section we will use (6.19), (6.21) and (6.23) to find the constitutive relationships needed for the plate reinforcement model introduced in Section 6.2.

6.4 Homogenization of the plate model microstructure

The homogenized stiffness coefficients of the reinforced plate model are now derived. The homogenization process is applied at the level of the elasticity matrix before integrating it in the thickness direction. Therefore, we make use of the results of the solution of the homogenization equation for plane stress conditions, as explained in Chapter 3.

In the case of plate model with ribbed reinforcement of Figure 6.1 the elasticity matrix for the isotropic solid, main plate layer is

$$\mathbf{C}^{(0)} = \frac{E_0}{1-\nu_0^2}\begin{bmatrix} 1 & \nu_0 & 0 \\ \nu_0 & 1 & 0 \\ 0 & 0 & (1-\nu_0)/2 \end{bmatrix} \tag{6.28}$$

where E_0 and ν_0 are respectively the Young's modulus and Poisson's ratio of the solid material of the plate. For the homogenized, orthotropic reinforcement material the elasticity matrix may be written as

$$\mathbf{C}^{(1)} = \frac{E_1}{1-\nu_1^2}\begin{bmatrix} d_{11}^h & d_{12}^h & 0 \\ d_{12}^h & d_{22}^h & 0 \\ 0 & 0 & d_{66}^h \end{bmatrix} \tag{6.29}$$

where E_1 and ν_1 are Young's modulus and Poisson's ratio of the solid material of the reinforcement, respectively. d_{ij}^h are the scaled homogenized coefficients of the assumed cellular microstructure for the reinforcement layers and are functions of the defining parameters of the microstructures, e.g. a and b in the microcell material model with rectangular holes.

The homogenized membrane stiffness matrix $\mathbf{D}_m^H$ can be found using (6.21). For the plate model of Figure 6.1 the components of $\mathbf{D}_m^H$ are

$$D_{ij\,m}^H = \int_{-t_0/2-t_1/2}^{-t_0/2} C_{ij}^{(1)}dz + \int_{-t_0/2}^{t_0/2} C_{ij}^{(0)}dz + \int_{t_0/2}^{t_0/2+t_1/2} C_{ij}^{(1)}dz \qquad i,j=1,2,6 \tag{6.30}$$

or

$$D_{mij}^H = t_0 C_{ij}^{(0)} + t_1 C_{ij}^{(1)} \qquad i,j=1,2,6 \tag{6.31}$$

where superscripts (0) and (1) are for the main plate and the homogenized reinforcement layer, respectively. For an isotropic plate with orthotropic reinforcement layers (6.31) can be written as

$$\mathbf{D}_m^H = \begin{bmatrix} \frac{E_0 t_0}{1-\nu_0^2} + \frac{E_1 t_1}{1-\nu_1^2} d_{11}^h & \frac{\nu_0 E_0 t_0}{1-\nu_0^2} + \frac{E_1 t_1}{1-\nu_1^2} d_{12}^h & 0 \\ \frac{\nu_0 E_0 t_0}{1-\nu_0^2} + \frac{E_1 t_1}{1-\nu_1^2} d_{12}^h & \frac{E_0 t_0}{1-\nu_0^2} + \frac{E_1 t_1}{1-\nu_1^2} d_{22}^h & 0 \\ 0 & 0 & \frac{E_0 t_0}{2(1+\nu)} + \frac{E_1 t_1}{1-\nu_1^2} d_{66}^h \end{bmatrix}. \tag{6.32}$$

By making use of (6.19) the homogenized flexural stiffness matrix $\mathbf{D}_f^H$ is calculated as follows

$$D_{ij\,f}^H = \int_{-t_0/2-t_1/2}^{-t_0/2} C_{ij}^{(1)} z^2 dz + \int_{-t_0/2}^{t_0/2} C_{ij}^{(0)} z^2 dz + \int_{t_0/2}^{t_0/2+t_1/2} C_{ij}^{(1)} z^2 dz \quad {}_{i,j=1,2,6} \tag{6.33}$$

which results in

$$\mathbf{D}_f^H = \begin{bmatrix} C_0\bar{t}_0 + C_1\bar{t}_1 d_{11}^h & C_0\bar{t}_0\nu_0 + C_1\bar{t}_1 d_{12}^h & 0 \\ C_0\bar{t}_0\nu_0 + C_1\bar{t}_1 d_{12}^h & C_0\bar{t}_0 + C_1\bar{t}_1 d_{22}^h & 0 \\ 0 & 0 & C_0\bar{t}_0(1-\nu_0)/2 + C_1\bar{t}_1 d_{66}^h \end{bmatrix} \tag{6.34}$$

where

$$C_0 = \frac{E_0}{12\left(1-\nu_0^2\right)} \quad , \quad C_1 = \frac{E_1}{12\left(1-\nu_1^2\right)} \quad , \tag{6.35}$$

$$\bar{t}_0 = t_0^3 \quad \text{and} \quad \bar{t}_1 = t_1\left(3t_0^2 + 3t_0t_1 + t_1^2\right). \tag{6.36}$$

The transverse shear matrix $\mathbf{D}_s$ can be obtained from (6.23). However, it is important to notice that the homogenization properties derived from the two dimensional elasticity problem do not provide any means for computing the transverse shear properties. Consequently, to determine these coefficients the homogenization equations need to be solved in the general three dimensional case. A typical approach followed in practice [10] is to consider the transverse shear matrix as

$$\mathbf{D}_s = \frac{1}{\epsilon}\begin{bmatrix} 1 & 0 \\ 0 & 1 \end{bmatrix} \qquad 0 < \epsilon \ll 1. \tag{6.37}$$

In this approach the transverse shear strain energy is viewed as a penalty contribution in the strain energy. Note the by letting $\epsilon \to 0$ the Kirchhoff's assumption (i.e. $\gamma = 0$) is enforced.

Here, in an attempt to take into account the transverse shear resistance of the homogenized reinforcement material, we make an analogy with the isotropic material which leads us to

$$D_{44}^H = \frac{\kappa E_0 t_0}{2(1+\nu_0)} + \frac{\kappa E_1 t_1}{2(1+\nu_1)} d_{22}^h, \tag{6.38}$$

$$D_{45}^H = D_{54}^H = 0\,, \tag{6.39}$$

$$D_{55}^H = \frac{\kappa E_0 t_0}{2(1+\nu_0)} + \frac{\kappa E_1 t_1}{2(1+\nu_1)} d_{11}^h. \tag{6.40}$$

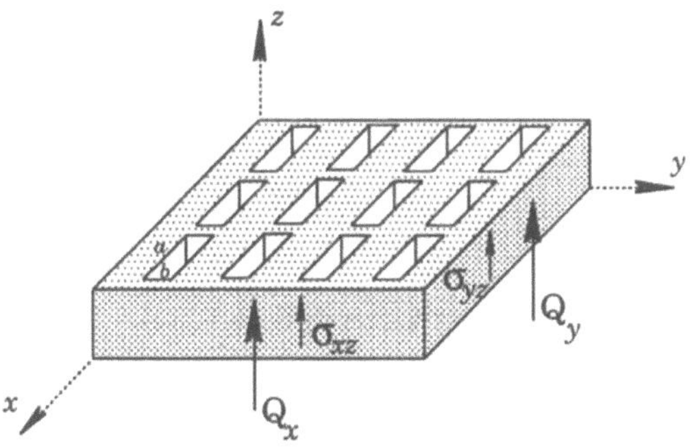

Figure 6.4 A group of cells of reinforcement layer.

It should be noted that d_{22}^h and d_{11}^h in (6.38) and (6.40), respectively, are chosen intuitively. Figure (6.4) tries to illustrate the reasoning of this approach.

With $a > b$, as in Figure 6.4, we know that $d_{11}^h > d_{22}^h$. It is also understandable that there are more shear resistance in planes parallel to zy than planes parallel to xz plane. Consequently, we could say $C_{55} > C_{44}$. Therefore, it is plausible to use d_{11}^h in place of Young's modulus for the shear coefficient C_{55} and to use d_{22}^h for C_{44}.

6.5 Optimization problem

As mentioned earlier, our objective is to find the stiffest plate with a certain amount of material used as reinforcement. Therefore, the optimization problem, for the sake of generality, may be expressed as the maximization of the total potential energy. By using the total potential energy as the objective function, one is also able to solve problems with prescribed boundary displacements without the need to change the formulation [13]. Here the formulation is derived for the material model with rectangular voids. Because of the close similarity there is no need to repeat it for other material models.

The total potential energy of the Mindlin-Reissner plate can be written as

$$\Pi(\mathbf{u}) = \frac{1}{2}\int_{\Omega_0} \Big[\Big(N_x\epsilon_x^0 + N_y\epsilon_y^0 + N_{xy}\epsilon_{xy}^0\Big) + \big(M_x\chi_x + M_y\chi_y + M_{xy}\chi_{xy}\big)$$
$$+ \big(Q_x\gamma_{xz} + Q_y\gamma_{yz}\big)\Big]\, d\Omega - \int_{\Omega_0} \mathbf{u}^{0^T}\mathbf{q}\, d\Omega, \tag{6.41}$$

where Ω_0 is the mid-surface area and $\mathbf{q}$ is a general force vector comprising gravity, transverse and in-plane components. The vector $\mathbf{u}$ contains the displacement field components. It is assumed that all forces are imposed on Ω_0 only. Equation (6.41) may also be written as

$$\Pi(\mathbf{u}) = \frac{1}{2}\int_{\Omega_0} \left(\mathbf{N}^T\boldsymbol{\epsilon}^0 + \mathbf{M}^T\boldsymbol{\chi} + \mathbf{Q}^T\boldsymbol{\gamma}\right) d\Omega - \int_{\Omega_0} \mathbf{u}^{0^T}\mathbf{q}\, d\Omega. \tag{6.42}$$

Using the finite element discretization and employing the constitutive relationships between generalized stresses and strains, we obtain

$$\Pi(\mathbf{u}) = \frac{1}{2}\sum_{e=1}^{N}\left[\int_{\Omega_0^e}\left(\epsilon^{0^T}\mathbf{D}_m\epsilon^0 + \chi^T\mathbf{D}_f\chi + \gamma^T\mathbf{D}_s\gamma\right)d\Omega - \int_{\Omega_0^e}\mathbf{u}^{0^T}\mathbf{q}\,d\Omega\right]. \tag{6.43}$$

The optimization problem can be expressed as

$$\begin{aligned} &\max_{a,b,\theta} && \Pi(\mathbf{u}) \\ &\text{subject to} && t_1\int_{\Omega_0}(1-ab)d\Omega \le \bar{V}_s \\ &\text{and} && 0 \le a, b \le 1 \end{aligned} \tag{6.44}$$

where t_1 is the total thickness of the reinforcement and $\bar{V}_s$ is the given volume of reinforcement. The displacement $\mathbf{u}$ satisfies the equilibrium equation.

Applying the finite element discretization, reminiscent of the procedure used for the plane stress problem, the Lagrangian function may be written as

$$\mathcal{L} = \Pi(\mathbf{u}) - \Lambda\left[t_1\sum_{e=1}^{N}(1-a^e b^e)\Omega^e - \bar{V}_s\right] - \sum_{e=1}^{N}\lambda_{a0}^e(-a^e) - \sum_{e=1}^{N}\lambda_{a1}^e(a^e-1)$$
$$- \sum_{e=1}^{N}\lambda_{b0}^e(-b^e) - \sum_{e=1}^{N}\lambda_{b1}^e(b^e-1). \tag{6.45}$$

From stationary of $\mathcal{L}$ in terms of the design variables (a^e, b^e, θ^e) and making use of the Kuhn-Tucker conditions, the optimality conditions similar to (4.63)-(4.70) are obtained. Adopting an optimality criteria method and following the same approach as that used for the plane problem a resizing scheme very similar to the plane problem (e.g. (4.81) and (4.82)) may be constructed.

It should be noted that using (6.27), the strain energy density in the plate problem can be written as

$$\epsilon^T\mathbf{D}\epsilon = \epsilon^{0^T}\mathbf{D}_m\epsilon^0 + \chi^T\mathbf{D}_f\chi + \gamma^T\mathbf{D}_s\gamma. \tag{6.46}$$

Therefore, in the plate bending problem (4.71) will be replaced by

$$E_a^e = \frac{1}{2}\frac{\int_{\Omega^e}\epsilon^{0^T}\partial\mathbf{D}_m^e/\partial a^e\epsilon^0 d\Omega + \int_{\Omega^e}\chi^T\partial\mathbf{D}_f^e/\partial a^e\chi d\Omega + \int_{\Omega^e}\gamma^T\partial\mathbf{D}_s^e/\partial a^e\gamma d\Omega}{-\Lambda t_1 b^e\Omega^e}. \tag{6.47}$$

Similarly, E_b can be obtained

$$E_b^e = \frac{1}{2}\frac{\int_{\Omega^e}\epsilon^{0^T}\partial\mathbf{D}_m^e/\partial b^e\epsilon^0 d\Omega + \int_{\Omega^e}\chi^T\partial\mathbf{D}_f^e/\partial b^e\chi d\Omega + \int_{\Omega^e}\gamma^T\frac{\partial\mathbf{D}_s^e}{\partial b^e}\gamma d\Omega}{-\Lambda t_1 a^e\Omega^e}. \tag{6.48}$$

The algorithm employed is then exactly the same as for the plane problem.

6.6 The finite element model

The requirement of only C^0 continuity of the displacements and independent rotations by the Mindlin-Reissner formulation allows the use of a wide range of interpolation schemes [18-20]. However, when exact numerical integration is used with standard Mindlin-Reissner finite elements, very disappointing results are obtained in applications to thin plates [14]. This phenomenon is called *locking* and it is believed that the shear strain energy is responsible.

A remedy to this problem is *selective integration* procedures in which a reduced integration rule is used to evaluate the stiffness matrix associated with the troublesome shear strain energy. As discussed before, the shear strain energy terms may be interpreted as penalty functions which force the shear strains to equal zero as the plate thickness is reduced. The imposition of these constraints leads to a deterioration of stiffness matrix and over-stiff results are obtained. By the use of the selective integration procedure, the over-constraining effect of the shear part of the stiffness matrix is alleviated . On the other hand, the use of reduced integration may produce the further problem of rank deficiency of the stiffness matrix. Consequently, in this procedure reduced integration is used for the shear portion of the stiffness matrix, but the remaining terms are fully integrated in an attempt to attain the required rank of the overall stiffness matrix [14, 18].

The ideal Mindlin-Reissner plate element should have the following features: the element

1. should not lock in thin plate situations,
2. should not have spurious zero energy modes,
3. should satisfy the usual completeness and convergence requirements,
4. should not be sensitive to element distortions,
5. should be easy to implement.

Bearing in mind the above requirements and based on the research results on such issues, the 9-node *heterosis* element developed by Hughes *et al* [16] is a good choice. The heterosis element is formulated using the 9-node Lagrangian shape functions for the rotations (θ_x and θ_y) and the 8-node serendipity shape functions for lateral displacement w. This element has correct rank and does not lock in thin plate situations [14].

To implement the heterosis 8-node serendipity and the 9-node Lagrangian elements it is more convenient to adopt a *hierarchical* formulation to represent all degrees of freedom [21]. Therefore, for a typical element the serendipity shape functions are adopted for nodes 1 to 8 and for node 9 the bubble function is used [14]. Thus, the hierarchical degrees of freedom at node 9 are perturbations from the associated serendipity interpolants. We can also use the hierarchical formulation for geometrical representation as well as for interpolating displacements.

In practice, to obtain the heterosis element representation, the hierarchical transversal displacement at node 9 is restrained to zero, but the independent rotation degrees of freedom are left free. The total displacements at node 9 are calculated using the expression

$$\mathbf{u} = \sum_{i=1}^{8} N_i^e(0,0)\mathbf{a}_i^e + \mathbf{a}_9^e \tag{6.49}$$

where $\mathbf{a}_9^e = \left(w_9 = 0, \theta_x, \theta_y\right)^T$.

6.7 Optimal rotation

Stationarity of the Lagrangian $\mathcal{L}$ in (6.45) with respect to the orientational design variable θ^e results in

$$\int_{\Omega^e} \epsilon^{0T} \frac{\partial \mathbf{D}_m^e}{\partial \theta^e} \epsilon^0 d\Omega + \int_{\Omega^e} \chi^T \frac{\partial \mathbf{D}^e}{\partial \theta^e} \chi d\Omega + \int_{\Omega^e} \gamma^T \frac{\partial \mathbf{D}_s^e}{\partial \theta^e} \gamma d\Omega = 0\,. \tag{6.50}$$

To find the updated orientational variable θ_{k+1}^e in each iteration k we may proceed as following:

1. Having updated a^e and b^e the homogenized elasticity matrix $\mathbf{D}^H\left(a^e, b^e\right)$ is determined.
2. Employing the frame rotation formula (4.85), $\mathbf{D}^G\left(a^e, b^e, \theta^e\right)$, $\mathbf{D}_s^G\left(a^e, b^e, \theta^e\right)$ and $\mathbf{D}_m^G\left(a^e, b^e, \theta^e\right)$ are determined as functions of the orientational variable θ^e.
3. Using (4.87), the sensitivities of the constitutive matrices with respect to θ^e can be constructed.
4. Having completed the analysis, the values of the generalized displacements have already been determined at the Gauss points. Consequently numerical integration of the mutual energies yields a trigonometric polynomial similar to (4.88).
5. This trigonometric polynomial can be solved by iterative methods such as the Newton-Raphson method.

In the absence of membrane forces, the angle of the principal moments may be used to update θ^e in a manner similar to the plane stress problem. It is noted that the principal directions of the plane layers parallel to the mid-plane are the same as the principal moments. This is because, according to the assumption that normals to the mid-surface remain straight after deformation, the stresses σ_x, σ_y and σ_{xy} in different layers of a plane are linearly proportional to the bending moments M_x, M_y and M_{xy}. Also note that the effect of the rotation angle on the transverse shear strength is negligible.

6.8 Examples

Based on the method explained in this chapter the plate version of the PLATO code was developed and given the name PLATO-PL. In this section the method will be demonstrated by running PLATO-PL for different material models.

For the examples presented below, the plate model of Figure 6.1 is adopted. This is a plate model with external reinforcement ribs. Unless otherwise stated the modified resizing scheme is used with the following parameters: move limit $\zeta = 0.008$ and volume tolerance $\delta = 0.001$. For the artificial material model, the penalty exponent is taken as $\mu = 3.2$. In all the examples below, the Young's modulus and Poisson's ratio for the plate, as well as the reinforcement layers, are taken to be $E = 100000.0$ and $\nu = 0.3$.

6.8.1 Simply supported square plate with a central point load

In this example the geometry and loading are shown in Figure 6.5. The intensity of the point load applied at the centre of the plate is $P = 10000$. All dimensions are assumed to be consistent. The thickness of the main plate and the external reinforcement layers are $t_0 = t_1 = 0.01$. See Figure 6.1. The area fraction for the reinforcement is set to $\Omega_{\text{solid}}/\Omega = 50\%$.

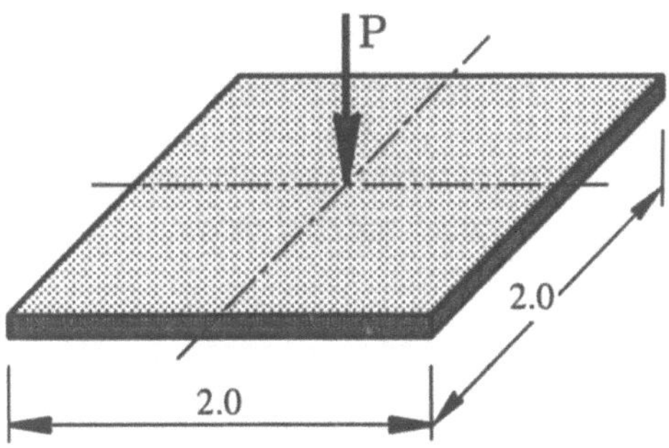

Figure 6.5 Square plate with a centric point load.

This problem was solved by dividing the whole domain into a mesh of 20 × 20 square elements. Using the bi-cubic version of the material model with rectangular holes, the emergence of the optimal layout is shown in Figure 6.6. The dark regions in the density plots represent the stiffened areas occupied by reinforcement. In the initial structure the reinforcement mass is assumed to be uniformly distributed.

The strain energy of the initial structure is 24.482 with a share of the shear strain energy of 0.049 (i.e. 0.20%). After 160 iterations the total strain energy has dropped to 13.392. The flexural strain energy is 13.346 while the share of the shear strain energy is 0.045 (i.e. 0.34%).

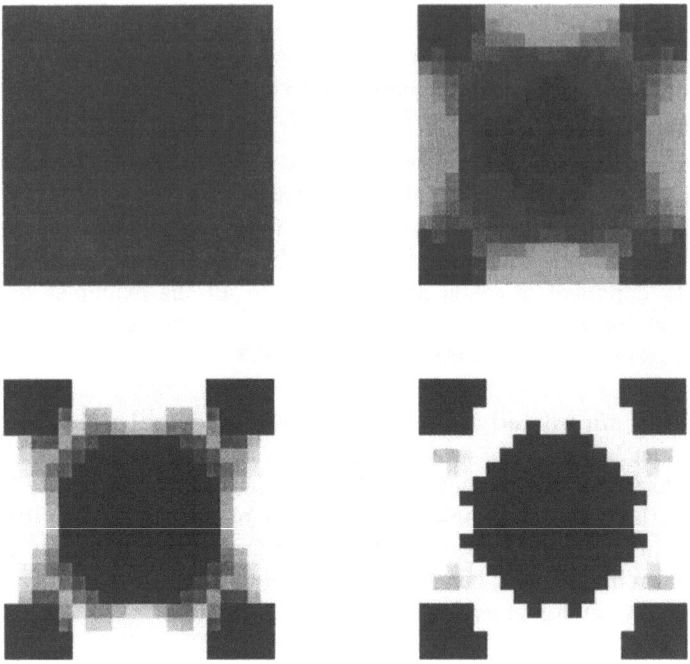

Figure 6.6 *Emergence of the optimal reinforcement layout for the simply supported square plate with a central point load using the bi-cubic material model.*

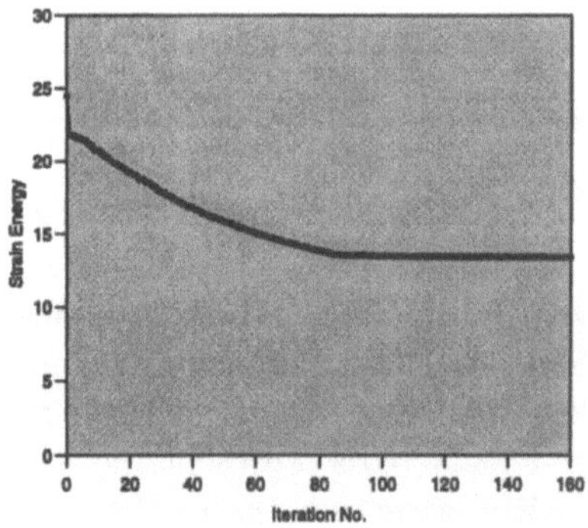

Figure 6.7 *Iteration history for the bi-cubic material model.*

The graph of the iteration history is illustrated in Figure 6.7. It is observed that after alignment of the major reinforcement material direction with the direction

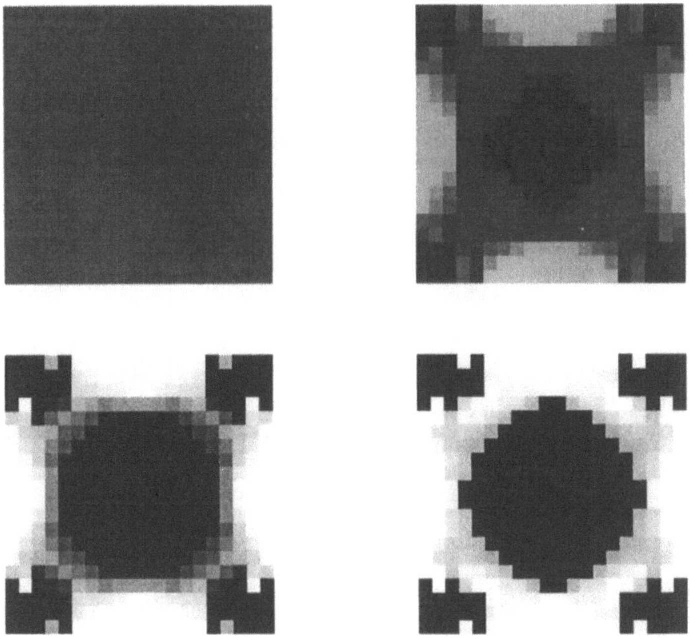

Figure 6.8 Emergence of the optimal layout using the bi-quartic material model.

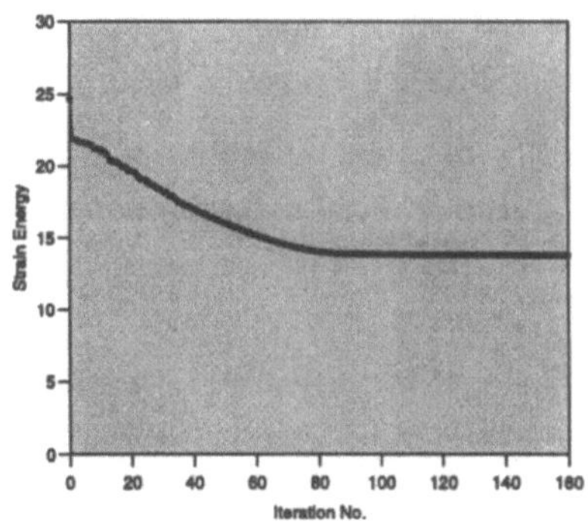

Figure 6.9 Iteration history for the bi-cubic material model.

of the principal moments (iteration 0), the strain energy drops from 24.481 to 21.854, where the share of the flexural and shear strain energies are: 24.433, 0.049 and 21.794, 0.059, respectively. This indicates that by correcting the orientation of the reinforcement material, the shear strain energy has increased slightly.

When the bi-quartic version of the material model with rectangular holes is used a slightly different layout is obtained. The resulting layout and the iteration history are illustrated in Figures 6.8 and 6.9 respectively. The convergence history is very similar to the bi-cubic case. The initial and final strain energy values in this case are 24.657 and 13.806, respectively, which are slightly higher than for the previous case.

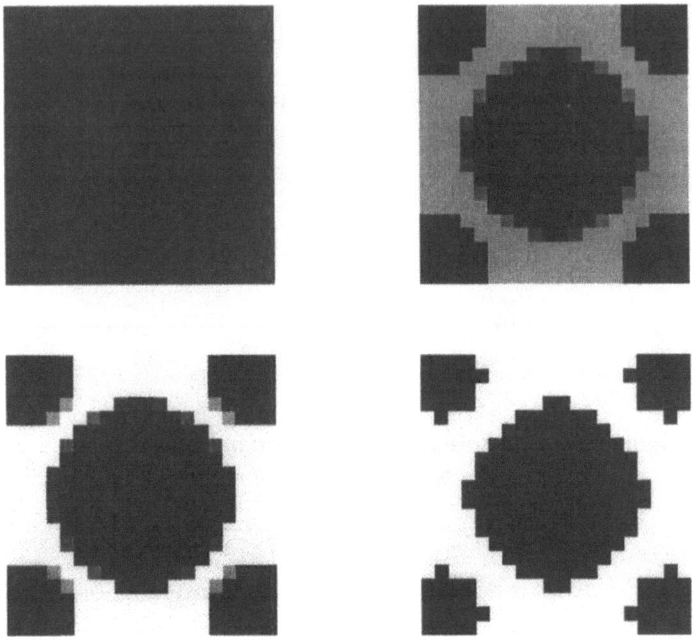

Figure 6.10 **Emergence of the optimal layout using the artificial material model.**

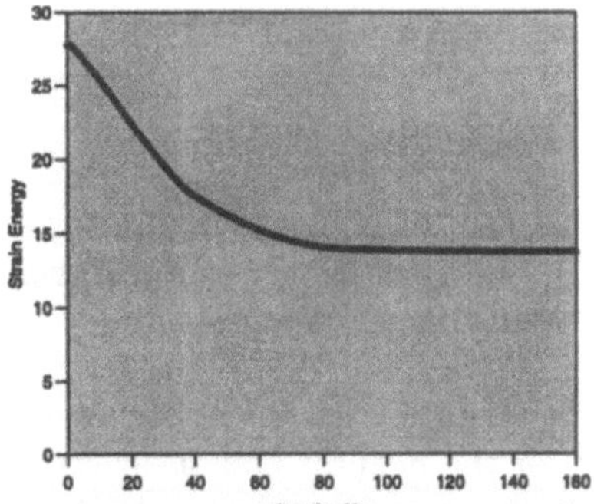

Figure 6.11 *Iteration history for the artificial material model.*

Using the artificial material model, the resulting layout is depicted in Figure 6.10. The convergence history is shown in Figure 6.11. The 'artificial' strain energy value for the obtained layout is equal to 13.752. As discussed in Chapter 5 this value should not be directly compared with the optimal values of

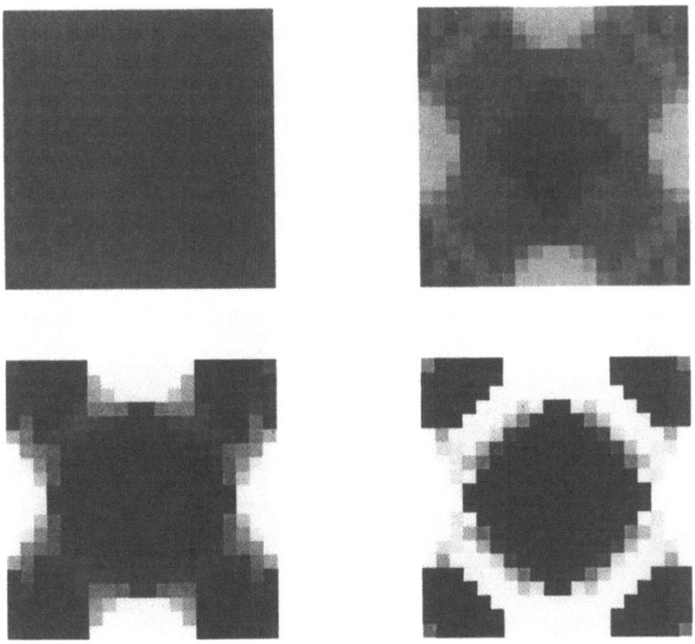

Figure 6.12 *History of emerging the optimal layout using the rank-2 material model.*

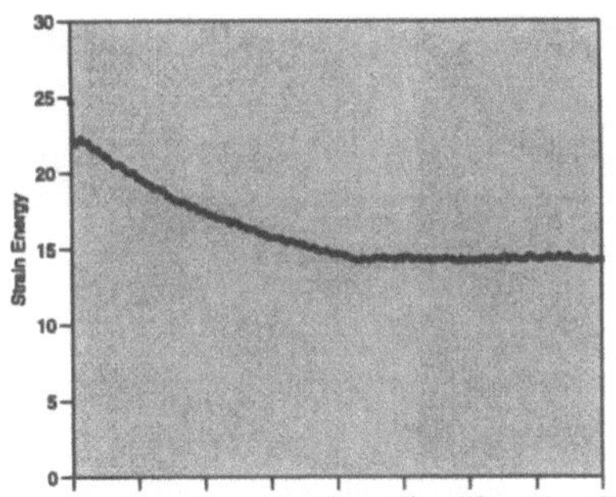

Figure 6.13 *Iteration history for the rank-2 material model.*

the earlier cases. The equivalent strain energy using homogenization and with the bi-cubic complete polynomial is 13.357. See Appendix 3.C. When the bi-quartic complete polynomial is used, the equivalent value is 13.768.

Finally, using the rank-2 material model the resulting layout and the convergence history are illustrated in Figures 6.12 and 6.13, respectively.

6.8.2 Simply supported square plate subject to a uniform load

The problem of the last section is now considered but with the loading changed to a uniformly distributed transverse load. All the other parameters are taken to be identical to those of Section 6.8.1. The finite element mesh adopted is as before. The reinforcement layouts obtained using different material models are illustrated in Figure 6.14. It is observed that the resulting reinforcement

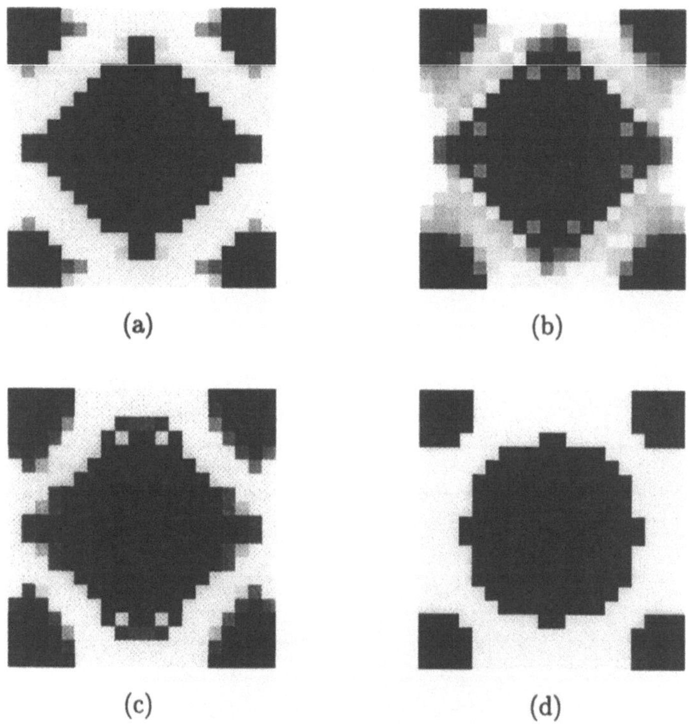

Figure 6.14 *Resulting layouts for the uniform transverse loading using different material models: (a) bi-cubic, (b) bi-quartic, (c) rank-2 and (d) artificial. Problem is solved for the whole domain without making use of symmetry.*

layouts for the uniformly distributed load case are to a large extent, similar to those obtained for the case with a central point load.

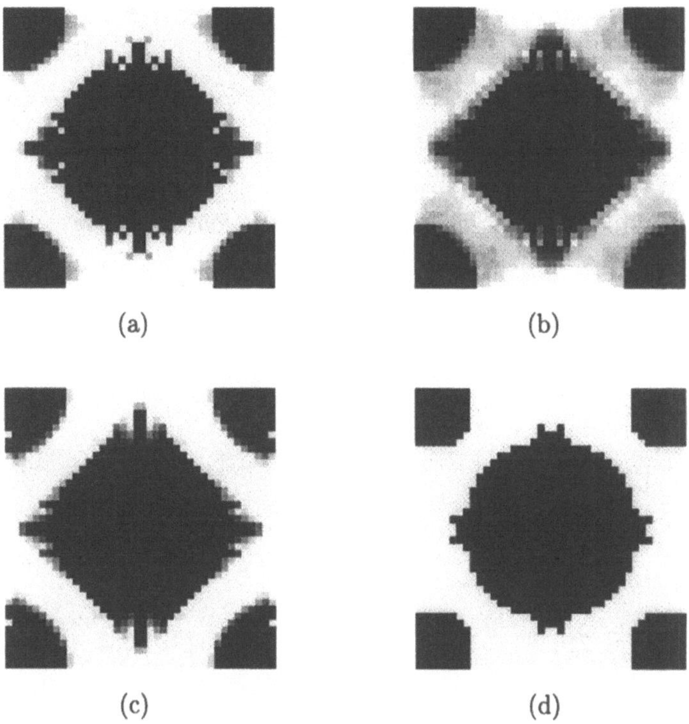

Figure 6.15 *Resulting layouts using symmetry and different material models: (a) bi-cubic, (b) bi-quartic, (c) rank-2 and (d) artificial.*

To study the effect of the mesh discretization on the resulting layout, use is now made of the bi-symmetry condition. Only a quarter of the plate is discretized using a mesh of 20×20 elements. Consequently, the mesh is four times finer. Results obtained using the different material models are given in Figure 6.15.

6.8.3 Square plate subject to four point loads

The simply supported plate problem is again considered with a loading layout as shown in Figure 6.16. The point loads are located symmetrically in the centre of the quadrants.

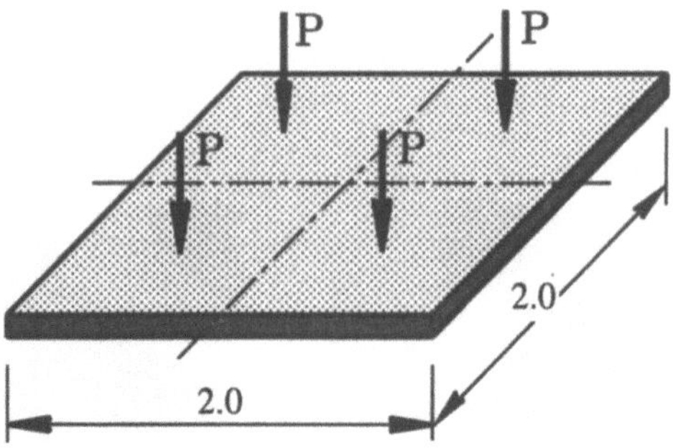

Figure 6.16 Square plate subject to four transverse point loads.

This problem was solved using the bi-symmetry condition and different material models. The resulting layouts are illustrated in Figure 6.17.

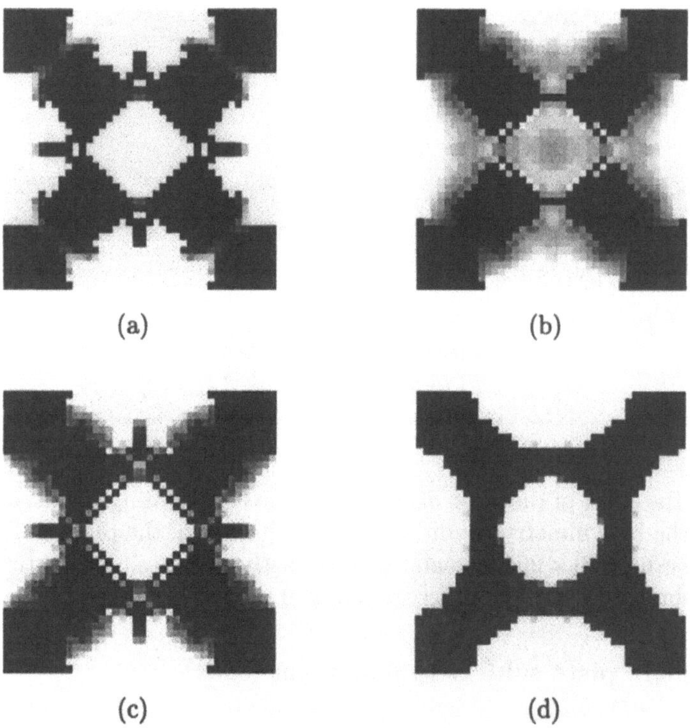

Figure 6.17 Resulting layouts using symmetry and different material models: (a) bi-cubic, (b) bi-quartic, (c) rank-2 and (d) artificial.

6.8.4 Square slab with a circular hole

The problem of finding the optimal layout of a simply supported square plate

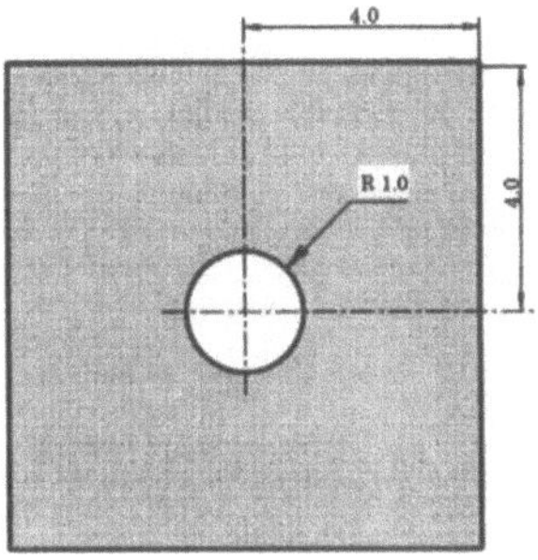

Figure 6.18 *Geometry definition of the square slab problem.*

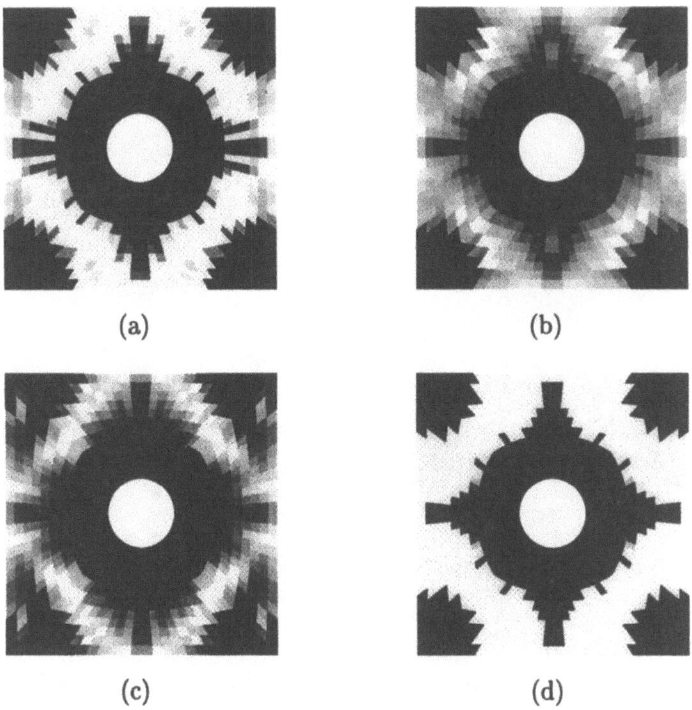

Figure 6.19 *Resulting layouts using symmetry and different material models: (a) bi-cubic, (b) bi-quartic, (c) rank-2 and (d) artificial.*

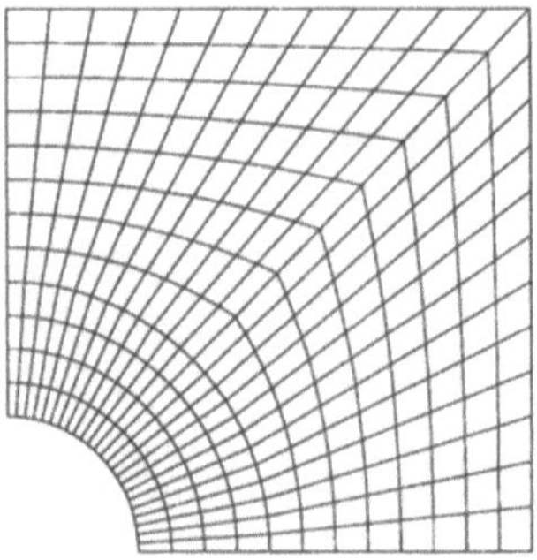

Figure 6.20 *Finite element mesh.*

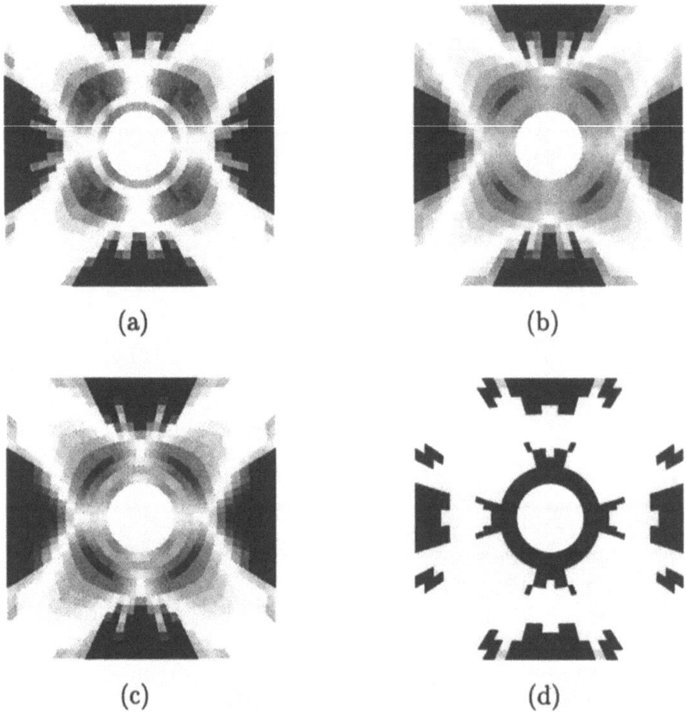

Figure 6.21 *Resulting layouts for the fully clamped supports using symmetry and different material models: (a) bi-cubic, (b) bi-quartic, (c) rank-2 and (d) artificial. The area fraction for the reinforcement is* $\Omega_{\text{solid}}/\Omega = 30\%$.

with a central circular hole is now considered. The plate is subject to a uniformly distributed load. The thickness of the main plate is $t_0 = 0.01$ and the thickness of the external reinforcement layers is $t_1 = 0.03$. The geometry is illustrated in Figures 6.18.

By making use of the bi-symmetry condition, this problem was solved only for a quarter of the domain. The assumed area fraction for the reinforcement is $\Omega_{\text{solid}}/\Omega = 50\%$. The modified resizing scheme was used with the parameters: $\zeta = 0.008$, $\delta = 0.001$ and $\mu = 3.2$. The resulting layouts using different material models are illustrated in Figure 6.19. The finite element mesh employed is depicted in Figure 6.20.

This problem was also solved for fully clamped supports on the straight edges. The given area fraction for the reinforcement is $\Omega_{\text{solid}}/\Omega = 30\%$. Using the same resizing parameters as before and the modified resizing scheme, the resulting layouts are shown in Figure 6.21. The finite element mesh adopted is the same as in Figure 6.20.

6.8.5 Flat slab of a multi-span floor

The problem definition of a middle two way flat slab is illustrated in Figure 6.22. The dimensions of columns and the span are given as 0.2 and 2.0 in both directions, respectively. The loading is assumed to be uniformly distributed over all bays with an intensity equal to 40000. The Young's modulus and the Poisson ratio are taken as 2.0×10^{11} and 0.3, respectively. The thicknesses of the main middle plate and the reinforcement layers are equal to 0.08. All dimensions are assumed to be consistent. The given area fraction for the reinforcement is $\Omega_{\text{solid}}/\Omega = 50\%$.

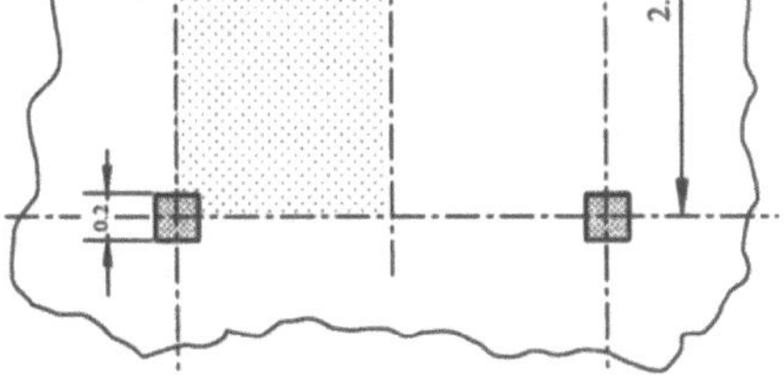

Figure 6.22 *Problem definition of the flat slab problem.*

Figure 6.23 Optimal layout for the reinforcement.

By making use of the bi-symmetry this problem was solved for a quadrant of a floor as is shown in Figure 6.22. Assuming that we have the same situation in the adjacent bays, the rotations and the vertical displacements of the columns are assumed to be zero. The resulting layout is illustrated in Figure 6.23. The artificial material model is used with the penalty exponent $\mu = 3.2$. The resizing parameters are: $\zeta = 0.008$ and $\delta = 0.001$.

6.9 Further developments

The formulation presented in this chapter is general and it can easily be extended to the reinforcement and layout optimization of shell problems which have a very wide field of application.

It should be noted that the homogenized membrane constitutive matrix (6.32) can be applied to the reinforcement optimization of plane problems as well. It is interesting to note that this approach can be further extended to the problems of vibration and buckling of plates.

Extension of the topology optimization technique to three dimensional problems provides a valuable tool for a design engineer to solve practical problems. Implementation of the method using a standard three dimensional finite element code is quite straightforward, especially when the artificial material models are used. Providing the capability to deal with the multiple load cases and the optimization of frequency response is very important for real life design problems.

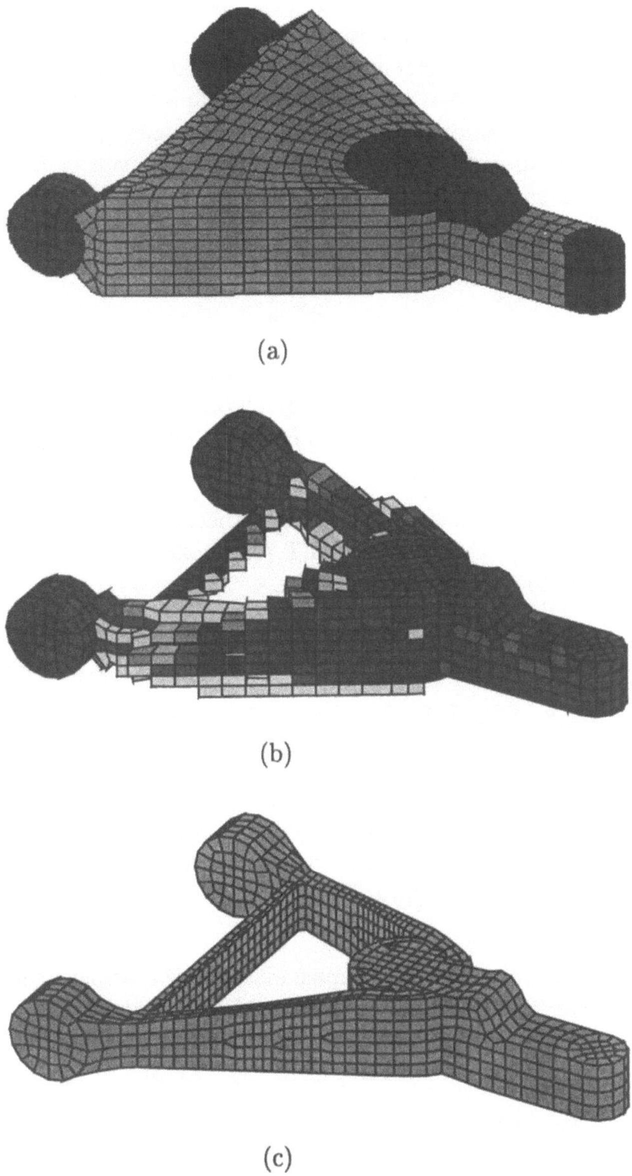

Figure 6.24 Control arm design. (a) Design and non-design domains, (b) density plot of elements and (c) Finite element model. Courtesy of Altair Computing Inc., Troy, Michigan, USA.

As an example of the use of the topology optimization method in practice, optimal design of a control arm by using a commercial software is illustrated in Figure 6.24. The design and non-design domains of the problem are depicted in Figure 6.24(a). The darker elements represent the non-design domain which are the attachment points for the frame, shock, spring seat stabilizer and spindle. Using the topology optimization technique and masking out elements with material densities below 30% the resulting layout is given in Figure 6.24(b). Using this layout as a guide, the finite element model for stress analysis is illustrated in Figure 6.24(c).

References

[1] Olhoff N., On singularities, local optima and formulation of stiffeners in optimal design of plates, in *Optimization in structural design*, edited by Sawczuk A. and Mroz Z., pp. 82–103. Springer-Verlag, New York (1975)

[2] Lurie K.A. and Cherkaev A.V., Prager theorem application to optimal design of thin plates, *MTT (Mecnanics of solids)*, **11**, 157–159 (1976)

[3] Cheng K.T. and Olhoff N., An investigation concerning optimal design of solid elastic plates, *Int. J. Solids Struct.*, **16**, 305–323 (1982)

[4] Rozvany G.I.N., Olhoff N., Bendsøe M.P., Cheng K.T. and Taylor J.E., On the solid plate paradox in structural optimization, *J. Struct. Mech.*, **10**(1), 1–32 (1982)

[5] Bendsøe M.P., Some smear-out models for integrally stiffened plates with applications to optimal design, in *In proceedings of Int. Symp. Optimum Structural Design*, pp. 13–34, Univ. Ariz., Tucson Arizona (1982)

[6] Bourgeat A. and Tapiero R., Homogenization of a transversely perforated plate in the frame of Hencky Mindlin's theory, in the thermoelastic case with non-uniform oscillating coefficients, *C. R. Acad. Sci. Paris, Serie I*, **297**, 213–216 (1983)

[7] Suzuki K. and Kikuchi N., Generalized layout optimization of three-dimensional shell structures, in *Geometric aspects of industrial design*, edited by Komkov V. SIAM (1991)

[8] Tenek L.H. and Hagiwara I., Optimization of material distribution within isotropic and anisotropic plates using homogenization, *Comp. Meth. Appl. Mech. Engrg.*, **109**, 155–167 (1993)

[9] Tenek L.H. and Hagiwara I., Optimal rectangular plate and shallow shell topologies using thickness distribution or homogenization, *Comp. Meth. Appl. Mech. Engrg.*, **115**, 111–124 (1994)

[10] Soto C.A. and Díaz A.R., Optimum layout and shape of plate structures using homogenization, in *Topology design of structures*, edited by Bendsøe M.P. and Mota Soares C.A., pp. 407–420. Kluwer Academic Publishers (1993)

[11] Díaz A.R., Lipton R. and Soto C.A., A new formulation of the problem of optimum reinforcement of Reissner-Mindlin plates, Private communications

[12] Kikuchi N. and Suzuki K., Structural optimization of a linearly elastic structure using the homogenization method, in *Shape and Layout Optimization in Structural Design*, edited by Rozvany G. (1990)

[13] Bendsøe M.P., Díaz A.R. and Kikuchi N., Topology and generalized layout optimization of elastic structures, in *Topology design of structures*, edited by Bendsøe M.P. and Mota Soares C. A., pp. 159–205. Kluwer Academic Publishers (1993)

[14] Hinton E. and Abdel Rahman H.H., Mindlin plate finite elements, in *Finite element software for plates and shells*, edited by Hinton E. and Owen D.R.J., pp. 157–233. Pineridge Press, Swansea (1984)

[15] Hughes T.J.R., Cohen M. and Haroun M., Reduced and selective integration techniques in the finite element analysis of plates, *Nuclear Engineering and Design*, **46**, 203–222 (1978)

[16] Hughes T.J.R. and Cohen M., The 'heterosis' family of plate finite elements, in *Proceedings of the ASCE Electronic Computations Conference*, St. Louis, Missouri (1979)

[17] Bendsøe M.P., *Optimization of structural topology, shape, and material.* Springer, Berlin (1995)

[18] Hughes T.J.R., *The finite element method.* Prentice-Hall International, Inc., Englewood Cliffs, New Jersey (1987)

[19] Hinton E. and Owen D.R.J., *Finite Element Programming.* Academic Press, London (1977)

[20] Zienkiewicz O.C. and Taylor R.L., *The finite element method*, vol. 2. McGraw-Hill, London, 4th edn. (1989)

[21] Zienkiewicz O.C. and Taylor R.L., *The finite element method*, vol. 1. McGraw-Hill, London, 4th edn. (1989)

Part III

Other Methods and Integrated Structural Optimization

7

ALTERNATIVE APPROACHES TO STRUCTURAL TOPOLOGY OPTIMIZATION

This chapter is devoted to intuitive methods which are simple to implement and may be used as an alternative to topology optimization by the homogenization method. A method for simulation of functional adaptation of bone mineralization in vertebrates is introduced and an algorithm based on effective stresses is presented. The evolutionary fully stressed method is also briefly explained.

7.1 Introduction

For the sake of completeness in this chapter we deal with some alternative approaches to the problem of topology optimization. These methods are intuitive and they suffer from the lack of a rigorous mathematical base. However, in comparison with the homogenization method, these methods are simple to implement and less costly. For certain problems they may yield very near optimum results. In practice, these methods can be employed to obtain a rough guide to the optimum layout which may later be used as an initial design for more accurate methods. However, it is important to be aware of the objectives and limitations of these approaches.

The simulation of natural processes has filtered into various disciplines of engineering control, design and optimization. As examples of these procedures, genetic algorithms, simulation of brain functions (neural networks), simulation of adaptive growth of trees and mineralization of bones can be named.

The main idea behind all natural process simulation methods is that the products of natural processes are, in some sense, optimal and furthermore, the way nature achieves this is itself an evolutionary optimization process. Although it is not always easy to prove this claim, because of the inherent constraints in natural life (e.g. the limitation of nutrients, and the inevitable competition between the different species), there is a higher probability of the survival of the better and fitter creatures and systems. Thus methods based on copying the characteristic functions of natural processes as evolutionary optimization techniques sounds plausible.

In accepting this idea, it should be noted that the optimization task in nature is a multi-objective one. Because of interaction with other phenomena in nature, true mimicry of a natural process is too complicated and almost impossible. However, copying certain rules and functions of nature has its uses.

Another intuitive approach to the structural topology optimization problem is the evolutionary fully stressed method. We will see later that there are strong similarities between this method and natural processes. The basic idea behind the evolutionary fully stressed method is that in a structure, material which has only a small contribution in carrying the load, i.e. only has small stress levels, is used inefficiently and therefore can be removed. This method will be briefly explained later.

7.2 Simulation of functional adaptation of bone mineralization

Bone is the main biological load carrying structural element in the bodies of vertebrates. Unlike most of the normal mechanical materials such as steel or concrete, bone has a mechanism that responds to mechanical stimuli by changing its shape and internal architecture. Some classes of bone cells are capable of adapting themselves, in a steady process of bone resorption and deposition, to a quasistatic equilibrium state, which reflects the mechanical loading environment to which a given bone is exposed [1].

In the regions of the bone material where the stress level is low, resorption occurs, whereas in highly stressed zones a deposition process of minerals takes place. As a result of this functional adaptivity the more highly stressed areas are denser and stiffer and in the bone material of these areas a higher degree of mineralization is observed.

The biochemical and biomechanical process of the functional adaptivity of bone is still not fully understood. However, several attempts have been made to construct a simple mathematical model for the relationship between the mechanical stimuli and bone growth and local degree of mineralization [2-4].

The main application of the simulation of bone function is in the field of biomechanics, for example, the prediction of the consequences of the implantation of prostheses as in orthopaedic or dental surgery on the shape and structure of bone. This method has also been employed by Reiter *et al* [1, 5] and Walter *et al* [6] to find the optimum topology and design improvement of composite models.

In 'remodelling' algorithms, quantities such as the von Mises equivalent stress, the absolute value of the maximum principal stress or the effective strain energy density are used as scalar measures of the local stress level. Here, a remodelling scheme suggested by Reiter *et al* [1] is briefly introduced. This scheme is general and uses the effective strain energy density as a measure of the actual internal stress level.

7.2.1 A remodelling scheme based on effective strain energy density

In this model, the mechanical stimulus is the difference between the effective strain energy density u_e and a homeostatic strain energy density u_n which can be a specific value indicating a balanced state or an average value of u_e over the domain. The value of u_e in the case of bone remodelling is calculated by an appropriate superposition of the strain energy densities obtained from different load cases, weighted according to the load cycles to take into account the loading time histories experienced by a bone. The following formula is suggested by Carter *et al* [3]

$$u_e = \left(\sum_{i=1}^{n} w_i u_i^k \right)^{1/k} \tag{7.1}$$

where n is the number of loading cases, k is a weighting factor and w_i are defined as

$$w_i = n_i / \sum_{j=1}^{N} n_j. \tag{7.2}$$

Here, n_i is the number of load cycles in load case i and N is the total number of load cycles in the all of the load cases. The strain energy density u_i can be written as

$$u_i = \frac{1}{2}\epsilon_i^T \sigma_i \tag{7.3}$$

where ϵ_i and σ_i are the strain and stress tensors, respectively, for load case i.

In the case of bone remodelling, since the bone material is a two phase porous tissue, making the following correction, i.e. using u_a instead of u, better reflects the strain energy which is actually stored in the mineralized tissue, so that

$$u_a = \frac{\rho_m}{\rho_a} u \tag{7.4}$$

where ρ_m and ρ_a are the density of the mineralized tissue and the apparent density of the composite, respectively.

Now, we are ready to establish a remodelling scheme. The following formulation is employed by Reiter *et al* [1, 5]. If π is a parameter that defines the shape or material of the bone and the variations of this parameter $\Delta\pi$ could, for example, be the surface growth perpendicular to the surface or the change in apparent density, then

$$\Delta\pi = \begin{cases} \Delta\pi_{min} & \text{for} \quad \Delta\pi < \Delta\pi_{min} \\ \beta_1 \left[u_e - u_n(1-s)\right] & \text{for} \quad u_e < u_n(1-s) \\ 0 & \text{for} \quad u_n(1-s) \leq u_e \leq u_n(1+s) \\ \beta_2 \left[u_e - u_n(1-s)\right] & \text{for} \quad u_e > u_n(1+s) \\ \Delta\pi_{max} & \text{for} \quad \Delta\pi > \Delta\pi_{max} \,. \end{cases} \tag{7.5}$$

Note that β_1, β_2 and s are remodelling parameters which for bone remodelling have to be found from experiments. It is noted that in this scheme the deviation between u_e and u_n must exceed a certain threshold level to activate the scheme. In other words a 'lazy zone' around u_n is assumed. See Figure 7.1. The dependencies between π and the elasticity coefficients have to be calculated according to an appropriate micromechanical material model.

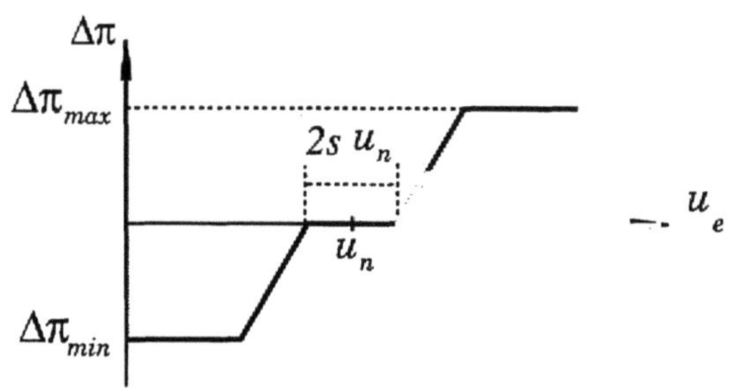

Figure 7.1 *Schematic relationship between u_e and the remodelling response.*

For simulation of bone remodelling, based on experimental investigations [2], the following relationship between the Young's modulus E and the apparent density ρ_a has been suggested

$$E = \alpha \rho_a^\gamma. \tag{7.6}$$

Parameters α and γ are determined by experiment and the Poisson's ratio is assumed to be constant. Assuming a foam-like microstructure for the bone, theoretical investigations predict γ values between 2 and 3 [7].

The above remodelling scheme combined with a finite element code leads to an iterative procedure for this adaptive remodelling algorithm. In each iteration the effective strain energy density in each element is determined using a finite element analysis. Consequently, by making use of the remodelling scheme, an adaptive model is obtained. Using this algorithm a good agreement with clinical observations was found for the simulation of the mass formation around a tibial implant system, with an axisymmetric finite element model and also a tooth implant, with a plane stress model [1, 5].

This remodelling algorithm has been implemented by Reiter *et al* [1, 5] to address the problem of optimum topology. They reported that in a few iterations a clear structural design was obtained. In fact, the convergence rate depends on the chosen growth-rate parameters. With reference to Pederson's work [8] they also argue that "for elastic structures the minimization of the overall compliance is equivalent to the local criterion of some optimal homeostatic u_n value" [1]. It should be noticed that this statement is valid only under the assumptions used in Pederson's work (e.g. the general optimality criterion of proportional gradients) [8, 9]. However, in Reference [5] they reported: "There does not exist any proof of uniqueness or optimality for these solutions". For structural topology, optimization based on the idea of uniform strain energy density, consideration of the lazy zone (see Figure 7.1) is probably not necessary.

7.2.2 A scheme based on effective stresses

In this scheme, a measure of the stresses is used instead of the strain energy density. The main idea behind this is the rule (axiom) of constant stress [10, 11]. According to this rule, biological load carriers in nature always adapt themselves towards a state of constant stress [12]. Therefore, in this approach the mechanical stimulus for the remodelling algorithm of bone mineralization is the level of stress. It is noted that with the adaptive method considered here, we gradually move towards the situation of uniform effective stresses or strain energy densities.

In this algorithm the modulus of elasticity E is chosen as the only parameter of bone remodelling and it is altered as a function of an indicator of the stress level. The Poisson's ratio ν is assumed to be constant.

As a measure of stress level, the absolute value of the maximum principal stress may be used. However, it is more common to use von Mises' equivalent stress as effective stress. The von Mises' equivalent stress is defined as

$$\sigma_{\text{eff}}^2 = \frac{1}{2}\left[(\sigma_1 - \sigma_2)^2 + (\sigma_2 - \sigma_3)^2 + (\sigma_3 - \sigma_1)^2\right] \tag{7.7}$$

where σ_i, $(i = 1, 2, 3)$, are the principal stresses. In two dimensional space (7.7) can be written as

$$\sigma_{\text{eff}}^2 = \sigma_x^2 + \sigma_y^2 - \sigma_x\sigma_y + 3\tau_{xy}^2 \ . \tag{7.8}$$

The algorithm of this procedure may be summarized as follows:

1. The first step is to define the design and non-design domains. The non-design domain is the part of the domain where the Young's modulus remains constant during the adaptation process. We also define loads, boundary conditions and material properties at this stage.
2. Next we generate a finite element mesh. We note that the mesh generated remains unchanged throughout the optimization process.
3. We then carry out a finite element analysis with an homogeneous Young's modulus for the whole design domain. The initial stress field is thus obtained.
4. Having determined the stresses, from (7.8) the effective stresses are calculated at the Gauss quadrature points of elements. To find a single value for the effective stress in each element the following weighted averaging formula may be used

$$\sigma_{\text{eff}}^e = \frac{\sum_{i=1}^{n_g} a_i \sigma_{\text{eff}}^{(i)}}{\sum_{i=1}^{n_g} a_i} \tag{7.9}$$

   where n_g is the number of Gauss points and a_i is the share of the Gauss point i in the volume (area) of the given element. Note that there are, of course, other approaches based on smoothing which could be used here. One may also use the value of the effective stresses at the centre of elements.
5. Having determined the effective stress values for different elements, we now vary the elasticity modulus of elements (E^e) according to a certain assumed relationship between the Young's modulus and effective stress,

$$E^e = f(\sigma_{\text{eff}}^e) \ . \tag{7.10}$$

   We note that various functions can be used for f in (7.10) to provide a 'direct' relationship between E^e and σ_{eff}^e. This direct relationship is necessary to simulate the adaptive mineralization of the bone where the areas under higher stresses become stiffer than areas with lower levels of stress. The issue of choosing an appropriate function f will be dealt with later.

6. Steps 3, 4 and 5 are repeated until a clear distinction between regions of high and low Young's moduli are found.
7. As the final step, a unique value of the elasticity coefficient is chosen and the corresponding contour is accepted to provide the boundaries of a new structural design proposal.

The flowchart of this algorithm is depicted in Figure 7.2.

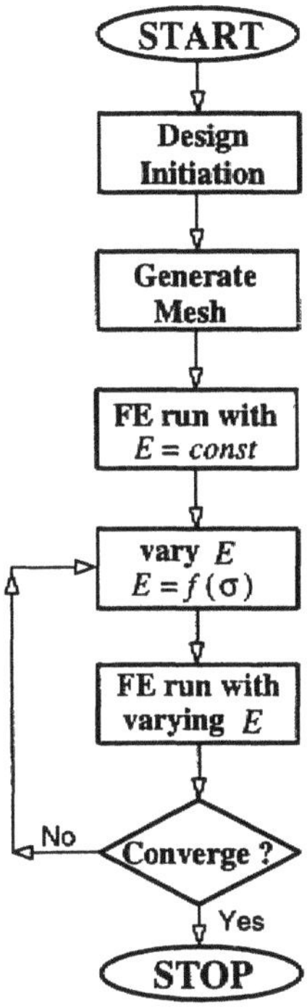

Figure 7.2 *Algorithm for topology optimization using the method of simulation of bone mineralization.*

As pointed out earlier, two questions to be answered are 'What is the proper function f in (7.10)?' or 'How do we vary the modulus of elasticity with regard to the calculated effective stresses?' Baumgartner *et al* [10] have investigated three different relationship as follows:

1. The stress relationship in the form

$$E^{(n+1)} = \sigma_{\text{eff}}^{(n)} \tag{7.11}$$

which means that in each iteration step the value obtained for the effective stress within a certain element is used as the modulus of elasticity in the next iteration. This procedure is very slow and in the case of complex examples yields poor results.

item2. The stress increment relationship of the form

$$E^{(n+1)} = E^{(n)} + k\left(\sigma_{\text{eff}}^{(n)} - \sigma_{\text{eff}}^{(n-1)}\right) \tag{7.12}$$

where the incremental change of the Young's modulus is proportional to the change of the effective stresses in the previous step. k is a magnification factor which increases the convergence speed of the scheme. Note that (7.12), in fact, is another form of

$$E^{(n+1)} = k\sigma_{\text{eff}}^{(n)} \tag{7.13}$$

and $k = 1$ yields the procedure described before.

In order to avoid a negative value for the Young's modulus and also to diminish the effect of stress concentrations, lower and upper bounds for the modulus of elasticity are considered. The lower bound of E is a very small positive number to prevent singularity in the stiffness matrix. This procedure offers a large variety of possible solutions with an increased number of iterations, particularly when the upper bound of the elasticity modulus is relatively low. This method seems to be well suited to aesthetical aspects or cheap manufacturing as it provides more choices.

3. The reference stress relationship where a global reference stress σ_{ref} is used instead of the effective stress from the previous iteration before. In place of (7.12), in this case, we have

$$E^{(n+1)} = E^{(n)} + k\left(\sigma_{\text{eff}}^{(n)} - \sigma_{\text{ref}}\right) \tag{7.14}$$

where σ_{ref} can be a constant given value or it may be taken as the value of the effective stress at a certain point such as a reference node. One also may take the average value of effective stresses over the whole domain as the reference stress σ_{ref}. In general, it is more efficient to start with a lower value for σ_{ref} and increase it in each step until a desirable value is achieved.

As with the previous case, we have to limit the Young's modulus, otherwise we will end up with a high stress concentration. For the maximum Young's modulus E_{max}, the desired value of the elasticity coefficient of the structure may be used. In the areas with no material (holes), for numerical stability, a minimum value $E_{min} = E_{max}/1000$ gives satisfactory results. In this procedure the convergence rate is higher than the cases before and it produces good results [10].

A linear relationship is assumed between the Young's modulus and effective stresses in the cases considered here. However, it is possible to use non-linear functions for this purpose.

As an example the SCB problem which was defined in Section 5.1, with a point tangent load applied at the mid point of the right hand side edge is here considered. See Figure 5.2. The following values were adopted: Young's modulus $E = 100000.0$, Poisson's ratio $\nu = 0.3$ and the intensity of the point load $P = 1000$. All dimensions are assumed to be consistent.

The domain of the problem is divided into a mesh of 640 equal, 4-node isoparametric elements. By employing the algorithm shown in Figure 7.2 and using the relationship (7.14) to update the modulus of elasticity, after only three iterations the contours of the resulting relative elasticity modulus E are illustrated in Figure 7.3. The convergence rate is very high even for values of k near to one and the final design is obtained after a very few iterations.

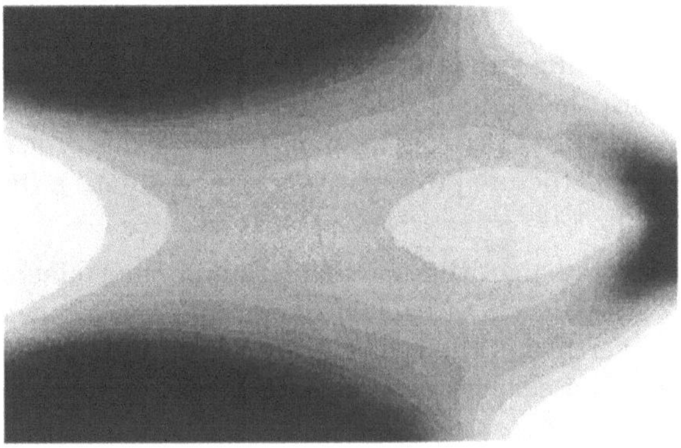

Figure 7.3 Contours of the Young's modulus.

The resulting 'optimal' layout may be obtained by extracting a structure boundary as an appropriate contour of the Young's modulus as shown in Figure 7.4.

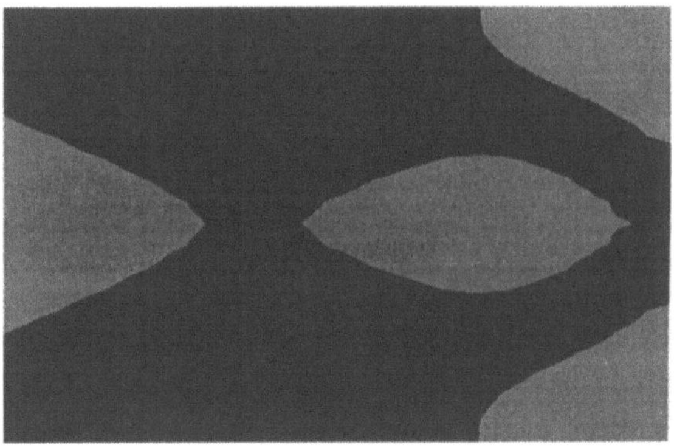

Figure 7.4 The resulting layout.

7.3 Evolutionary fully stressed design method

The fully stressed condition is an intuitive optimality criteria which has already been applied in the optimization of truss structures by several researchers (see for example Reference [13]). However, it has been shown that in general the optimum light-weight structure is not necessarily a fully stressed structure [14]. The reason is that for an indeterminate structure, it may happen that members which are not fully stressed help to relieve stresses in other members. A structure is considered to be fully stressed if all members (or elements) of that structure are either at their minimum limits or have reached their allowable stress levels under at least one of the design load conditions.

The extension of the fully stressed idea to two and three dimensional continua, in combination with the finite element method, provides a simple alternative approach to the structural topology optimization. In this procedure the lowly stressed regions of a structure are iteratively removed until a (quasi-)optimum state is reached [15].

It is noticeable that the algorithm for simulation of bone mineralization based on the effective stress level, as explained in Section 7.2.2, can also be used here. The only difference is the special form of the relationship between the Young's modulus of an element and the effective stress value, which here is a step function of the form

$$E^e = f\left(\sigma^e_{\text{eff}}\right) = \begin{cases} 0 & \text{if } \sigma^e_{\text{eff}} < \bar{\sigma} \\ E^0 & \text{if } \sigma^e_{\text{eff}} \geq \bar{\sigma} \end{cases} \tag{7.15}$$

where E^0 is the given elasticity modulus of the assumed material for the design domain, $\bar{\sigma}$ is a threshold effective stress value. Therefore, elements which have an effective stress value less than $\bar{\sigma}$ are removed (or killed). Hence, some researchers has given the name ***hard kill*** to this method [16]. The method explained in Section 7.2.2 is sometimes called the ***soft kill*** optimization (SKO) method, since instead of removing of the elements with low stresses, we make them weaker and softer (soft kill) [6, 10].

In practice, to avoid changing the finite element mesh during the optimization process, a very small positive value is assigned to the elasticity modulus E of the elements which are 'removed'. Therefore, (7.15) is altered to

$$E^e = \begin{cases} E_{min} & \text{if } \sigma^e_{\text{eff}} < \bar{\sigma} \\ E^0 & \text{if } \sigma^e_{\text{eff}} \geq \bar{\sigma} \end{cases} \tag{7.16}$$

where E_{min} is a very small specified number, chosen carefully so that it does not cause singularity in the stiffness matrix.

For a stable algorithm in the evolutionary fully stressed design method, it is crucial to allow only a small fraction of elements to be eliminated in each iteration. This can be controlled by careful specification of the threshold value $\bar{\sigma}$, which is assumed to be a fraction of the maximum value obtained for the effective stresses of the elements in the domain [15], so that

$$\bar{\sigma} = R_r \left(\sigma^e_{\text{eff}}\right)_{max} \tag{7.17}$$

here R_r is the *rejection ratio.* Often, after some element elimination cycles, a steady state condition is reached, where only a very few or no elements can be eliminated. Thus, the rejection ratio R_r has to be increased to enable the iteration to proceed further. The following relationship may be used for this purpose:

$$R_r^i = R_r^0 + (i-1)R_e \qquad (i = 1, 2, \cdots) \tag{7.18}$$

where i is the number of times that a stationary situation has occurred and R_r^0 is the *initial rejection ratio.* R_e is called the *initial rejection ratio.*

he method explained here is very easy to implement and any commercial finite element software can be used for this purpose, provided that it has an option of having different material properties and post processing facilities which allow the drawing the contours of effective stresses. After each analysis, the contours of the effective stresses are drawn and the modulus of elasticity of elements located in the low stress zone are changed to E_{min}.

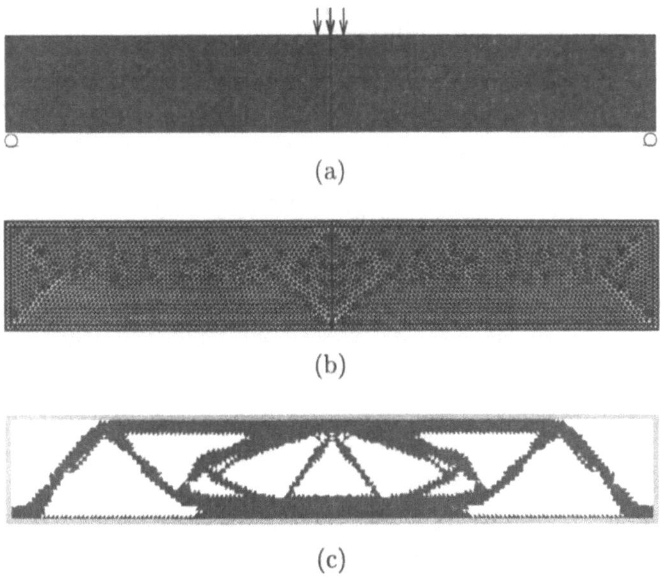

Figure 7.5 *The MBB beam. (a) Problem definition, (b) finite element mesh and (c) resulting topology. Courtesy of Dr J. Sienz, Civil Eng. Dept., University of Wales, Swansea, UK.*

As an example, the simply supported beam as shown in Figure 7.5(a) is considered. This beam, which is known as MBB, is part of structure of the floor of a civil aircraft. The finite element mesh is illustrated in Figure 7.5(b). For functionality reasons a surrounding non-design domain is defined around the beam. Employing the fully stressed evolutionary method, the resulting topology is illustrated in Figure 7.5(c).

The methods explained in this chapter can be easily employed for multiple load cases. In each iteration the maximum value obtained for the effective stress value of an element from all of the load cases, should be taken as σ^e_{eff}. Also note that the finite element analysis of a structure for multiple load cases is not much more expensive than for a single load case, as the global stiffness matrix remains unchanged and only the right hand side changes. This method can also easily be employed in free vibration and buckling problems. The interested reader is advised to consult Reference [17].

References

[1] Reiter T.J. and Rammerstorfer F.G., Simulation of natural adaptation of bone material and application in optimum composite design, in *IUTAM symposium on optimal design with advanced materials*, edited by Pederson P., pp. 25–35, Lyngby, Denmark (1993). Elsevier Science Publishers

[2] Carter D.R. and Hayes W.C., The compressive behaviour of bone as a two-phase porous structure, *J. Bone and Joint Surg.*, **59-A**, 954–962 (1977)

[3] Carter D.R. and Fyhrie D.P., Trabecular bone density and loading history: regulation of connective tissue biology by mechanical energy, *J. Biomech.*, **20**, 785–794 (1987)

[4] Huiskes R., Weinans H., Grootenboer H.J., Dalstra M., Fudala B. and Slooff T.J., Adaptive bone-remodeling theory applied to prosthetic-design analysis, *J. Biomech.*, **22**, 1135–1150 (1987)

[5] Reiter T.J., Rammerstorfer F.G. and Böhm H.J., Structural design improvement by functional adaptation, *Structural Optimization*, pp. 361–368 (1993)

[6] Walther F. and Mattheck C., Local stiffening and sustaining of shell and plate structures by SKO and CAO, *Structural Optimization*, pp. 181–188 (1993)

[7] Gibson L.J. and Ashby M.F., *Cellular solids - structure and properties*. Pergamon Press, Oxford (1988)

[8] Pedersen P., On thickness and orientational design of orthotropic materials, *Structural Optimization*, **3**, 69–78 (1991)

[9] Masur E.F., Optimum stiffness and strength of elastic structures, *J. Eng. Mech. Div. ASCE*, **EM5**, 621–649 (1970)

[**10**] Baumgartner A., Harzheim L. and Mattheck C., SKO soft kill option. The biological way to find an optimum structure topology., *Int. J. of Fatigue*, **14**(6), 387–393 (1992)

[**11**] Mattheck C., Engineering components grow like trees, *Mat.-wiss. u. Werkstofftech*, **21**, 143–168 (1990)

[**12**] Mattheck C., *Trees – The mechanical design.* Springer Verlag, Heidelberg, New York (1991)

[**13**] Haftka R.T. and Gürdal Z., *Elements of structural optimization*, chap. 9. Kluwer Academic Publishers, third edn. (1992)

[**14**] Razani R., Behavior of fully stressed design and its relationship to minimum weight design, *AIAA J.*, **3**(12), 2262–2268 (1965)

[**15**] Xie Y.M. and Steven G.P., Optimal design of multiple load case structures using an evolutionary procedure, *Engineering Computations*, **11**, 295–302 (1994)

[**16**] Hinton E. and Sienz J., Fully stressed topological design of structures using an evolutionary procedure, Tech. rep., University College of Swansea (1993)

[**17**] Xie Y.M. and Steven G.P., *Evolutionary structural optimization.* Springer-Verlog, Heidelberg (1997)

8

INTEGRATED STRUCTURAL OPTIMIZATION

An overview of the concept and modules of a three phase integrated structural optimization system is the subject of this Chapter. The topology optimization module provides information about the optimum layout and topology. In the image processing module by employing computer vision techniques a structural model with smooth boundaries is extracted. In the third phase, using the conventional size and shape optimization methods the final optimal design is obtained. By constructing integrated design and optimization systems considerable improvement may be achieved by cutting development time and design costs.

8.1 Introduction

Traditionally the structural design process starts with an initial conceptual model which is inspired by the designer's intuition and previous experiences. Subsequently analysis is carried out and the response of the structure is computed. Later, based on the calculated response a new design is suggested. The interactive process of 'design, calculation, new design' is continued until a satisfactory design is obtained. Further studies and tests on the functionality of the design as well as manufacturing and assembly requirements is carried out by simulation modelling and prototyping. The outcome of this stage is either a fully confirmed design to be passed for production or a new design to be analysed again.

In the very competitive markets of the last decade of the twentieth century, the traditional procedure of engineering design is too slow. It is very important for companies to cut development time and cost as well as to improve the quality of their products. Hence, concurrent or simultaneous engineering is inevitable to keep an edge over competitors rather than following the conventional sequential approach. The key success is a fully integrated process from the beginning of the design to the rapid production of the prototype in a computer aided design and manufacturing environment [1].

By employing optimization techniques in place of the traditional design process a considerable speed up and efficiency is obtained. In fact, by a fully integrated optimization system the nature of the design problem is reversed. That is, instead of suggesting a design and analysing it to see whether it fulfills the objectives and requirements, an optimization problem is constructed and based on the objectives and constraints, the unknown design is found. Optimization is a valuable tool to generate efficient innovative layouts for structures which go beyond the experience of the designer [2].

In the next section an overview of a three phase integrated design and optimization system is provided.

8.2 Overview of integrated structural optimization

As noted earlier, topological structural optimization methods can only give a rough idea about the boundary curves (surfaces) of a structure. These methods are carried out on a fixed domain without the need to change the initial finite element discretization and they result in an optimal material distribution layout which in some areas may be neither solid nor void (i.e. grey areas). Also, as has been noticed before, in topology optimization methods the material properties vary from element to element. The outcome of this procedure can be plotted as an equivalent density plot to produce a grey scale pattern representing the optimum layout. Because each element has its own equivalent density, the boundaries of the final layout have a 'jagged' form. A reasonable prediction of finer details of the boundaries of the optimum layout requires a very large number of finite elements in the discretized model which is too costly [3, 4].

To specify a mechanical body, usually the shape is defined by some segmental boundary curves. In conventional shape optimization methods the optimal form of these boundary curves is determined. Unlike topology optimization methods where only global objectives (such as mean compliance, average stress, weight, etc) can be dealt with, in shape optimization methods it is also possible to define local objective functions. In practice, design optimization often requires more considerations than global objectives. For example the objective function to be minimized might be the maximum von Mises' stress or the maximum value of the displacement with various stress, strain and displacement constraints or constraints due to the stability, manufacturability or even aesthetic requirements. Although the optimal layout obtained from topological optimization provides the engineer with useful information, to fulfill these requirements more precisely, the use of conventional shape optimization methods such as the boundary variation method is necessary.

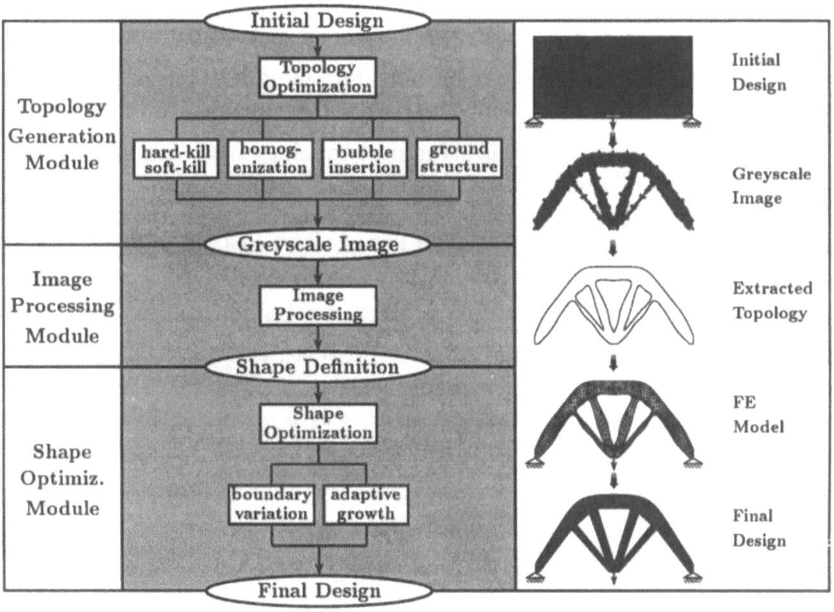

Figure 8.1 *Integrated structural optimization [2]. Courtesy of Dr J. Sienz, Dept. of Civil Eng., University of Wales, Swansea, UK.*

Consideration of the above points, leads us to try to construct an integrated design and optimization system with two main modules: topology optimization and shape optimization. Since these two approaches are fundamentally different, an interfacing module is required to identify and interpret patterns and to generate boundary curves from the output of the topology optimization module which is, as we have noted, a grey scale density plot. To do this, computer vision techniques such as pattern recognition, noise cancelling and curve smoothing may be employed. Computer vision is defined as the automated construction of explicit meaningful descriptions of physical objects from images, given in the form of density arrays [5, 6]. This middle phase may be called the *image processing* module. The procedure is illustrated in the right hand side of Figure 8.1. In the following, different modules will be explained in more detail.

8.3 Topology optimization module

As we have already seen the input to this module consist of the initial design domain (i.e. a reference domain which covers the space of interest), loading and boundary conditions and the specified volume of material. For functionality of the design some other global specifications may be considered. Based on these specifications the reference domain is divided into design and non-design domains. Non-design domains can comprise solid or void regions.

Usually the output of the topology optimization module is the array of optimal material density distribution, which when plotted yields a grey scale image.

The main concern of this book is structural topology optimization by the homogenization and related methods which are the most robust methods and were discussed in Chapters 1 to 6. Alternative intuitive procedures based on *simulation of bone mineralization (soft kill method)* and the *evolutionary fully stressed (hard kill)* method were the subject of Chapter 7. Also, the analytical approach to the layout optimization problem was introduced in Chapter 4.

During the last few years, some other approaches in structural optimization have evolved that the issue of topology is also dealt with. For the sake of completeness of presentation the *ground structure method* [3, 7-10] for trusses and other discrete structures and the *bubble method* [11, 12] for continua are briefly considered here.

8.3.1 Ground structure method

This method is suitable for truss or frame structures and in fact it is an integrated approach where topology, shape and size optimization are carried out simultaneously.

At the outset, a set of grid-like distributed points and their connections are assumed and all points or a selected subset of them are connected by truss members. This is called the *universe* or *ground structure*. By employing an optimality criteria procedure (e.g. DCOC method [13]), the optimal substructure of this structural universe is found. The method starts from the ground structure and removes certain members while resizing the remaining ones in an iterative process. This method results in truss-like structures. For mean compliance minimization the optimality criteria derived in Section 4.3.2 may be applied to this method. For more details the References [9, 10, 14, 15] should be consulted. Layout optimization of a clamped beam-like truss with a single vertical load (Example 1 in Section 4.8) is illustrated in Figure 8.2.

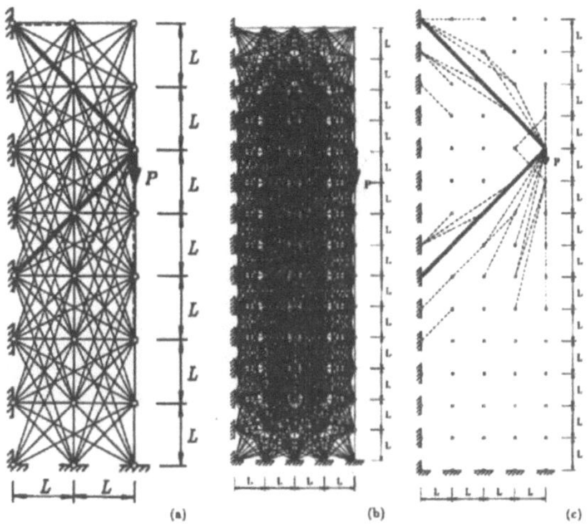

Figure 8.2 *Topology optimization of a clamped deep beam-like truss by the ground structure method. (a): a coarse ground structure with solution, (b): a fine ground structure, and (c): solution [8]. Courtesy of Professor G.I.N. Rozvany.*

8.3.2 Bubble method

This method is, in fact, an extension of the conventional boundary variation method for shape optimization, to also take into account the effect of topology. The basic idea is to iteratively position new holes (or bubbles) in a structure to change the topology class, followed by a hierarchically structured shape optimization to determine the optimal shape of the current bubble and external boundaries [12]. To describe the boundaries of bubbles and the external boundary of structure B-splines and Bezier curves may be used [16].

For optimization problems with simple global functionals (e.g. total potential energy and the volume of the structure), the optimum position of the bubble can be found analytically by variational methods. In general numerical methods should be applied, where one bubble in an arbitrary initial position is inserted and then it is 'guided' to the optimal position by means of mathematical programming or optimality criteria methods.

The optimization process can be divided into the following steps:

1. Starting from the initial design reference domain and considering the objective and constraint functions, conventional shape optimization with boundary variation method is carried out.
2. By inserting a new hole, the topology class is changed. Then either analytically or numerically, the optimum position of the bubble is found.
3. A shape optimization process is carried out in order to find the optimal shape of the new bubble and to determine the influence on the other variable boundaries.
4. Steps 2 and 3 are repeated until a convergence criterion† is satisfied.

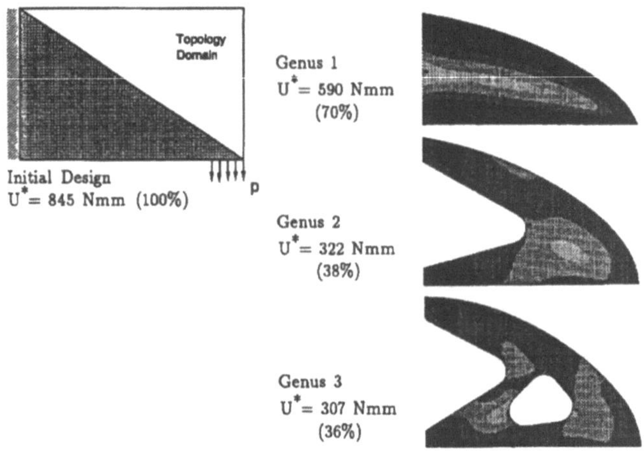

Figure 8.3 *Topology optimization of a clamped beam using the bubble method. (Top left): problem definition. (Top right): optimized boundaries for the initial structure. (Middle): after the first bubble insertion. (Bottom): final topology and shape after the second bubble insertion [11]. Courtesy of Professor H. A. Eschenauer*

In order to be able to provide, if necessary, sharp edges as well as smooth curved

† Note that this is not an easy task and still has to be developed [12].

ones, NURBS (non-uniform rational B-splines) are preferred [12]. Figure 8.3 shows the bubble method applied to the topology optimization of a clamped beam. The area occupied by the material stays the same throughout the optimization and is equal to the shaded area of the initial design. We note that the bubble method may be interpreted as an integrated structural optimization approach.

8.4 Image processing module

The outcome of the topology optimization module with the homogenization method (or the evolutionary fully stressed method) is an array of density of material in elements which from a design point of view is not entirely satisfactory. Apart from having 'jagged' boundaries, some elements may have intermediate densities (grey elements), especially when near optimal microstructures (i.e. rank-2 material model or cells with rectangular holes) are used†. Therefore, we are faced with the problem of extracting the boundaries of structure from the material density array which is represented by a contour diagram. This structure is used as an initial design for the conventional shape optimization module [17].

Depending on the value of the volume constraint, the resulting material distribution corresponds to an optimum topology that can vary from a truss-like configuration to a closed solid shape. Since discrete structures can be understood as a special case of continuum structures, here we focus on the latter case where in a two dimensional space the images generated in the topology optimization module can be modelled by plane stress or plane strain representations [18].

Chirehdast *et al* [19] used a thresholding procedure to convert the grey-scale image with densities between 0 and 1 to a binary image. That is, the densities lower than a given value are considered as void and above that value as solid. Then they applied a bitmap operation to remove noise from the binary image. Later, they use closed and open B-splines and linear segments to obtain smooth boundaries.

The selection of the control points of the boundary segments can be done either manually [2, 5] or automatically [18]. In a manual parameterization, the designer is free to remove the isolated zero density elements or to consider a number of adjacent voids as a single hole [17].

By considering the output layout of Example 2 in Section 4.8, a simple procedure based on density isolines is illustrated in Figure 8.4. The density isolines

† "Although with the present technology for producing advanced composite materials, one may argue that we should not limit ourselves to black and white designs only" [3].

are depicted in Figure 4.2(b). To do this it is only necessary to find the density values at nodal points by averaging the densities of adjacent elements and then use the conventional contour plotting technique. In Figure 8.4(c) using the threshold method the appropriate isoline is chosen. Finally by performing some curve smoothing, the finite element mesh is shown in Figure 8.4(d). Note that the definition of the boundary segments by splines is done manually, but we notice that here we do not have 'jagged' boundaries any more.

8.4.1 Elimination of mesh dependency and checkerboard problems using noise cleaning techniques

Introduction: Sigmund [20] suggested a method based on ideas borrowed from image processing techniques, to address the problems of checkerboard patterns and mesh dependency in layout optimization of linear elastic structures. Using this method, it is also possible to satisfy some manufacturability constraints such as exclusion of geometry change below a prescribed scale. He used the artificial material model and the so called *continuation* approach [21], where a variable penalty exponent (μ in (3.18))is used†.

The key point of this method is to regard the discretized design domain in topology optimization as a digital image. Each element is analogously considered as one pixel and its colour (density) is represented by a grey scale, where white is void and black is solid material.

Isolated small voids, a dense element surrounded by a lower density area or checkerboard patterns can be interpreted as unwanted noises and therefore, noise reduction techniques in digital signal processing [22] can be used to prevent them.

Several linear and nonlinear noise filtering techniques have been developed and may be employed in structural layout optimization problems. Although nonlinear techniques generally perform better than linear methods, they are not recommended for use in the layout optimization because they destroy the smoothness of the design problem [20]. Linear noise filtering techniques can be divided into two main approaches: *Fourier transformation* based techniques and *convolution* techniques. Fourier transformation based methods have the disadvantage that they can only be applied to regular rectangular meshes. Therefore, to be able to use these methods, we should either restrict ourselves to structured meshes with rectangular elements or we have to map the discretized domain into a regularly meshed domain with rectangular divisions. In other

† The continuation approach starts with a low value for $\mu = \mu^0$ and by allowing the subproblem to converge, increases its value by $\Delta\mu$. This process is continued until the desired value μ_{max} is reached. The values 1.4, 1.0 and 3.0 are suggested for $\mu = \mu^0$, $\Delta\mu$ and μ_{max}. Also, see Section 5.5.1.

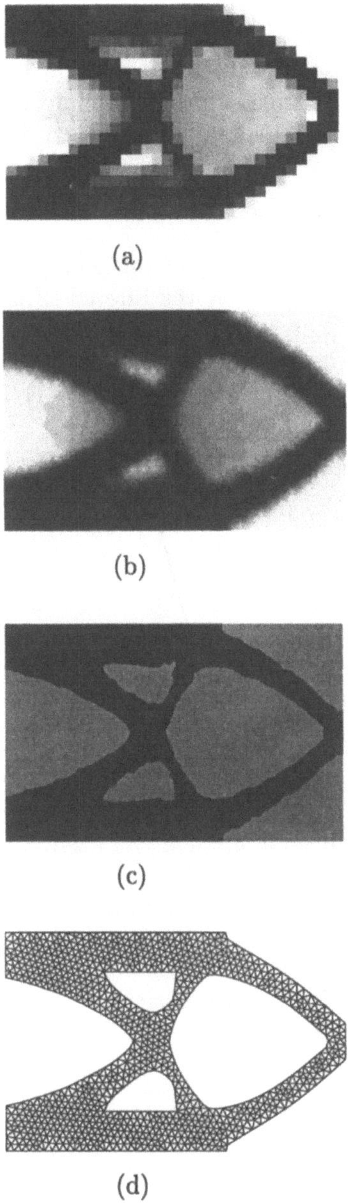

Figure 8.4 Image processing: (a) Output of topology optimization module, (b) Contours of density of material, (c) Choosing the appropriate contour and (d) Finite element mesh for the shape optimization module.

words, the design domain and the analysis domain should be separated. A similar procedure is followed in adaptive structural topology optimization [18].

Convolution based methods: In convolution based methods the colour (density) of each pixel is changed according to information from its neighbourhood (filter window). The convolution process can be formulated as

$$\rho^*(i,j) = \sum_{k=1}^{n}\sum_{\ell=1}^{n} \rho(k,\ell)\, H(i+m-k, j+m-\ell), \tag{8.1}$$

where $\rho(k,\ell)$ is the colour (density) of the pixel located in the kth row and ℓth column of the image and ρ^* is the filtered density of the pixel. The so called impulse response matrix $[H(i,j)]$ is an $n \times n$ square matrix that has to be chosen according to the purpose of the filter (e.g. image restoration or edge cleaning). The variable m is defined as $m = (n+1)/2$, where n is the number of pixels in each side of the filter window.

For noise cleaning the following low-pass form is widely used for the impulse response matrix:

$$\mathbf{H} = \frac{1}{(b+2)^2}\begin{bmatrix} 1 & b & 1 \\ b & b^2 & b \\ 1 & b & 1 \end{bmatrix}, \tag{8.2}$$

where the 'filter factor' $b \in [1, \infty]$. When $b = 1$ a simple averaging is performed and when b approaches infinity $\rho^*(i,j)$ approaches $\rho(i,j)$ and thus no filtering is provided. Note that here $n = 3$ and

$$\sum_{i=1}^{n}\sum_{j=1}^{n} H(i,j) = 1\,. \tag{8.3}$$

The convolution noise cleaning method can easily be applied to the result of the topology optimization process. Doing this, the rapidly varying densities will be filtered out and the checkerboard regions will be converted into 'grey' regions which are also undesirable in the final design. Also, it is not possible to apply the convolution procedure, as described above, inside the optimization process (i.e. to apply it after a few iterations), as it ruins the optimality conditions. Hence, Sigmund [20] suggested that a noise cleaning process should be integrated in the optimization problem.

In the case of using artificial material model, the optimization problem and the optimality condition can be written in terms of the strain energy density. In this case, use of a weighted average of strain energies over the neighbouring elements (filter window) is suggested by Sigmund. When quadrilateral elements are used, the relative strain energies may be modified as

$$\bar{w}_e = \frac{\sum_{i=1}^{n_H^e} H_i^* w_i}{\sum_{i=1}^{n_H^e} H_i^*} \tag{8.4}$$

where n_H^e (≤ 9) is the number of neighbouring elements for the element e and H^* is equal to H defined in (8.2) without the scaling factor. w_i is the strain energy of element i and $\bar{w}_e$ is the filtered strain energy of the element e.

For the isotropic artificial material model with the density of element $1-ab=\rho$ taken as design variable, (3.20) can be written as

$$\mathbf{D} = \rho^{\mu}\mathbf{D}^0 \tag{8.5}$$

where $\mathbf{D}^0$ is the elasticity matrix of an isotropic and homogeneous material and μ is the penalty exponent. Using (8.5) the strain energy of an element can be written as

$$w_e = \rho_e^{\mu} w_e^0 \tag{8.6a}$$

where the strain energy w_0^e is defined as

$$w_e^0 = \int_{\Omega^e} \mathbf{u}^T\mathbf{D}^0\mathbf{u}d\Omega \tag{8.6b}$$

where $\mathbf{u}$ is the displacement vector and Ω^e is area of element. By making use of (8.6a) the modification of the relative strain energy of elements as given by (8.4), may be written slightly differently as

$$\bar{w}_e = \frac{\sum\limits_{i=1}^{n_H^e} H_i^* w_i^0 \rho_i^{\mu}}{\rho_e^{\mu} \sum\limits_{i=1}^{n_H^e} H_i^*} \, . \tag{8.7}$$

It should be noted that by implementing this method, the original optimization problem is changed and as a result of seeking clear and checkerboard-free images, sub-optimal results for the value of the objective function are obtained. This fact is illustrated in Figure 8.5. Although, using lower values for the filter factor b in (8.2) results in clearer images with thicker bars, higher values for the strain energy W (to be minimized) are also obtained. The filter factor is chosen between 10 and 15 for checkerboard-free layouts [20].

This method is also applicable to the homogenization method where the mutual energy densities of elements are modified as above.

In the case of discretizing the design domain by triangular elements the impulse response matrix $\mathbf{H}$ in (8.2) will be replaced by a 2×2 matrix. For a three dimensional problem, having discretized with 8-node brick elements, the impulse matrix will contain 27 elements.

As noted in Chapter 5, the solution of the structural topological optimization problem is dependent on the employed finite element mesh especially when artificial material is used. For example, when a finer mesh is used, thinner structural elements (bars) and voids appear; See Section 5.5. From a practical point of view, we are usually more interested in solutions with a few large holes and a fixed topology regardless of the discretization.

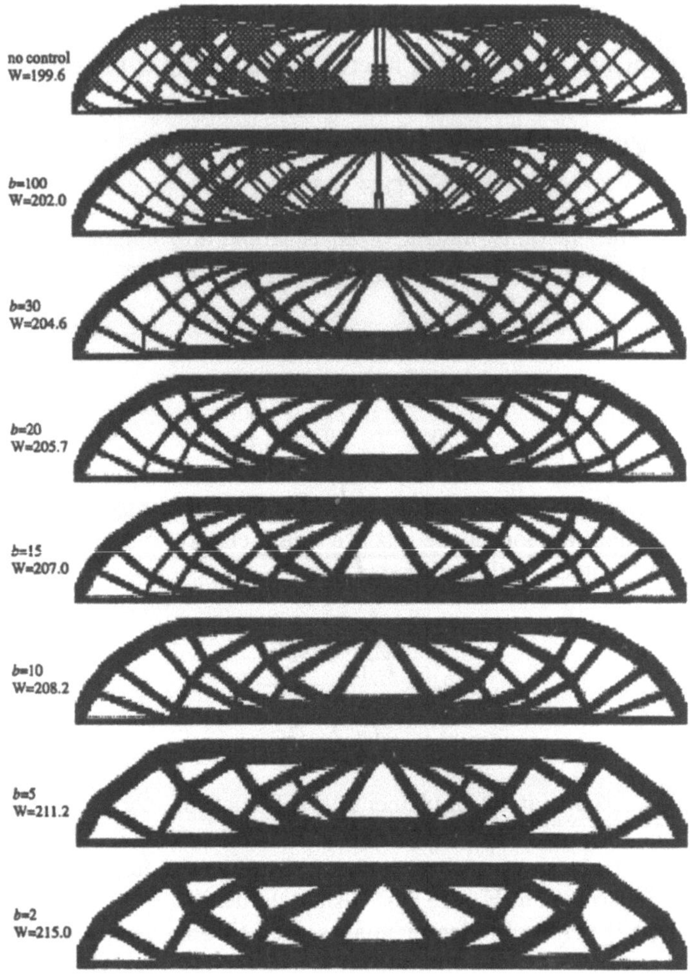

Figure 8.5 *Influence of the filter factor b on the optimal layout and the strain energy W [20]. Courtesy of Dr O. Sigmund, Technical University of Denmark, Lyngby, Denmark.*

Mesh dependency: To address the problem of mesh dependency Jog *et al* [23] proposed the perimeter control method. This method penalizes the intermediate density values, while adding a new nonlinear constraint, i.e. the perimeter of the structure. Because of strong dependency of the algorithm on the choice of penalty factors, convergence of this method is reported to be rather unstable [20].

To solve the problem of mesh dependency and in order to obtain more practical results, Sigmund has extended the convolution method to prevent the creation of elements with dimensions less than a given size r_{min} in the optimum layout. The convolution filter is here defined as

$$\hat{H}_i = a_i \left[r_{min} - r(e,i) \right] \quad i \in \{1, 2, \cdots, n_H^e\} \tag{8.8}$$

where a_i is the area (volume) of the element i and $r(e,i)$ is the distance between the centers of elements e and i. n_H^e is the the number of elements that satisfy $r(e,i) \leq r_{min}$.

Note that this method requires a search loop to find the neighbouring elements of a considered element. Although this search is computationally expensive, it only has to be done once at the beginning of the procedure.

In this case when the artificial material model is used, the strain energy density of an element e may be modified to

$$\hat{w}_e = \frac{\sum_{i=1}^{n_H^e} \hat{H}_i w_i}{\sum_{i=1}^{n_H^e} \hat{H}_i} \ . \tag{8.9}$$

Using this method, especially together with the continuation method, very interesting results were reported by Sigmund [20] which were mesh independent and checkerboard free. In the three left hand side pictures of Figure 8.6, the mesh dependency with the solutions using different number of elements with the filtering factor $b = 15$ is illustrated. Also, the effect of mesh refinement on the objective function, strain energy W, is given. On the right hand side of Figure 8.6 the results of the mesh independent algorithm are depicted. We observe that an increase of the number of elements does not change the topology of the structure. However it has the undesirable effect of increasing the value of the objective function.

Figure 8.7 shows another example of elimination of mesh dependency by application of the modified convolution filter. It is observed that having the same value for r_{min}, all solutions yield the same topology.

According to Sigmund choosing r_{min} to be 1.1 times the element size makes the modified filter very similar to the checkerboard filter with filter factor $b = 15$. Therefore, to prevent formation of checkerboards r_{min} should be chosen to be greater than or equal to 1.1 times the maximum element size. The resulting optimal topology can be further post-processed by using a method similar to that explained earlier; see Figure 8.4.

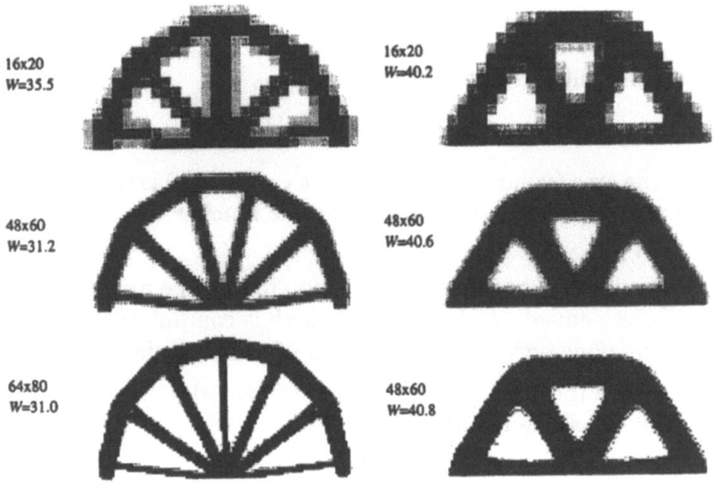

Figure 8.6 *Mesh dependency (solutions on the left) and elimination of mesh dependency (right) [20]. Courtesy of Dr O. Sigmund.*

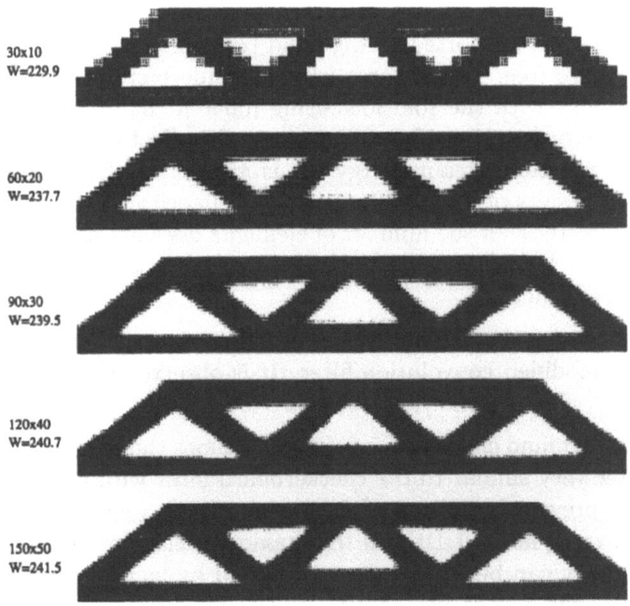

Figure 8.7 *Elimination of mesh dependency by using the modified convolution filter [20]. Courtesy of Dr O. Sigmund, Technical University of Denmark, Lyngby, Denmark.*

8.5 Shape optimization module

The shape optimization module can be seen as a method for post-processing of results from the topology optimization module. The output of the topology optimization module is an optimal layout with rough boundaries and some grey areas. In the image processing module the boundaries of the structure are extracted and preferably smoothed. It is also possible that the result of the topology optimization module be interpreted as a skeletal (truss-like) structure. This is suitable when the ratio of the material volume to the design domain volume is relatively low. An automatic process for extracting a truss structure is explained by Bremicker *et al* [5].

The shape optimization module leads us to an optimized detailed design where the additional design constraints are met. Geometrical restrictions associated with manufacturing and assembly requirements are considered and other performance criteria such as maximum displacement of a certain point or points, maximum stress or buckling of slender members are also taken into account. If necessary, sizing optimization may also be carried out at the same time. Here we focus on the general case of two dimensional continua.

Over the last two decades various methods have been suggested for structural shape optimization with an overall continual improvement. These methods vary from each other in respect of selection of design variables, parameterization of boundaries, finite element mesh generation, calculation of sensitivity gradients and solution techniques. In most of these methods mathematical programming methods coupled with finite element analysis is employed.

Optimality criteria methods can also be employed in shape optimization problems. Rigorous optimality criteria are derived mathematically, for example, by using variational methods. Intuitive optimality criteria may enforce the condition that the tangent stress or the strain energy density should be uniform along the boundary.

As the boundary element method only requires the modelling of the boundary, it does not have the difficulties of the finite element method in mesh regeneration and calculation of sensitivity derivatives stemming from inaccurate boundary representation. However, some erratic performance was reported by some researchers which has damaged the perception of the reliability of this approach [24, 25].

In pattern transformation methods, the shape of the boundary is transformed based on the stress level in the finite elements. First the stress ratio in the boundary finite elements is calculated and then the size of the boundary elements is scaled up or down, based on their stress ratio [26]. Instead of stress ratios strain energy ratios may also be considered [27].

Photoelasticity techniques can also be applied to optimize the shape of holes in plates and shells experimentally. Gradually material from low stress regions are removed to obtain uniform stress condition as well as significant reduction in the stress intensity factor. The numerical equivalent of this experimental approach has also been developed [28].

For more information one can consult the references provided by Haftka *et al* [24] in a survey paper on structural shape optimization.

Based on simulation of growth of trees in nature, the so called 'adaptive growth method' suggested by Mattheck [29] has recently been developed. This method has similarities with the above mentioned pattern transformation methods.

In the following, for the sake of completeness the well developed conventional boundary variation method is briefly reviewed. Then the adaptive growth method will be considered.

8.5.1 Boundary variation method

For a linear elastic body the aim of this method is to find, by means of boundary variation, the shape of the domain $\Omega \subset \mathbb{D}$ that minimizes (or maximizes) the objective function with respect to the state (equilibrium) as well as the behaviour and geometric constraints. The optimization problem may be expressed as

$$\begin{array}{llll} \min\limits_{\Omega \subset \mathbb{D}} & F(\mathbf{s}) & & \\ \text{subject to} & h_j(\mathbf{s}) = 0, & j = 1, 2, \cdots, n_h & \\ & g_k(\mathbf{s}) \leq 0, & k = 1, 2, \cdots, n_g & \end{array} \tag{8.10}$$

where $\mathbb{D} \subseteq \mathbb{R}^3$ denotes the set of admissible shapes defined through local geometric constraints (e.g. $s_i^\ell \leq s_i \leq s_i^u$). The design variables defining the shape are denoted by s_i, $(i = 1, \cdots, n)$. h_j and g_k are the equality and inequality constraints. The objective function $F(\mathbf{s})$ can be the volume, weight or strain energy.

Various mathematical programming techniques may be used to solve the optimization problem such as sequential quadratic programming (SQP) [30, 31] or the method of moving asymptotes (MMA) [32]. The algorithm for an automatic shape optimization process with adaptivity is illustrated in Figure 8.8.

The procedure consists of the following steps:

1. *Creation of geometrical design model and construction of optimization problem:* As the first step, the geometry of the structure under investigation is defined. It is important to define the geometry accurately and by as few parameters as possible. This is because the parameters defining the geometry are taken as the design variables of the shape optimization problem. In early work on shape optimization, the coordinates of the boundary nodes of the finite element model were used as design variables. This method showed serious drawbacks [24]. In some other efforts boundaries

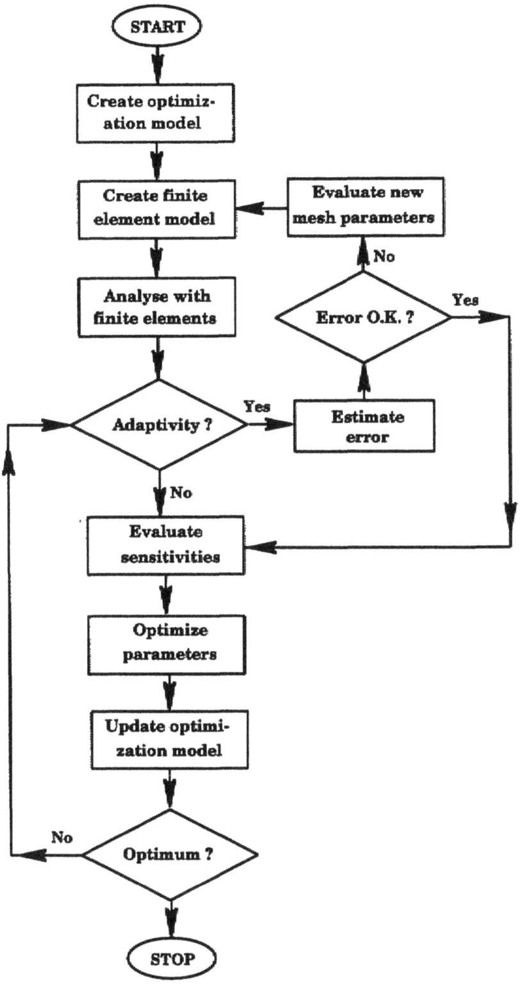

Figure 8.8 *Algorithm for structural shape optimization using boundary variation method with adaptivity [33].*

were represented by polynomials or linear combinations of shape functions with coefficients being the design variables. When high order polynomials are used, this method may result in an oscillatory boundary shape. The cubic spline functions, having two continuous derivatives everywhere and possessing minimum mean curvature are a very good choice for defining the moving boundary for two-dimensional problems.

The idea of separating the design model from the analysis model is an efficient method and provides more flexibility for the geometry description. The boundaries of the design domain (which can be divided into 'design elements' [24]) are described by using splines which, in turn, are defined by a set of key points. Some of the coordinates of these key points will be the design variables of the optimization problem. As the finite element analysis model varies during the optimization process, all data necessary to define the optimization problem (e.g. loading, boundary conditions, material properties and constraints) are associated with the geometrical design model only.

2. *Creation of finite element analysis model:* For analysis, an automatic mesh generator is usually used to discretize the domain to build a valid finite element model. The mesh generator needs to be able to transfer information about loading and boundary conditions from the geometrical design model to the finite element model. The use of an unstructured advancing front mesh generator is preferred to the macro-element mesh generators used previously in shape optimization.
3. *Analysis:* The displacements and stresses are evaluated together with the objective and constraint functions using the created finite element model and carrying out analysis.
4. *Adaptivity:* Since the shape of the domain varies during the optimization process, adaptive mesh refinement is needed in order to have a reasonable accuracy. This refinement can be obtained either by adding additional elements in the area to be refined (h method), or by increasing the order of the finite elements (p method). In the adaptive procedure we start with an initial mesh density and after carrying out the analysis, the discretization error is estimated by one of the recovery techniques (e.g. super-convergent patch recovery and Zienkiewicz-Zhu error estimator [34, 35]). If the error is satisfactory we continue, otherwise a new mesh density is evaluated based on the error distribution and a new mesh is generated. This procedure is repeated until the desired error is achieved.
5. *Sensitivity analysis:* In most of mathematical programming algorithms, to find a search direction, the sensitivity information is required. The derivatives of the objective and constraint functions with respect to the design variables are calculated using a sensitivity analysis, which indicates the sensitivities of the current design to small changes in the design variables. The sensitivities can be computed by finite difference, analytical or semi-analytical methods†.
6. *Solution of optimization problem:* Using the function values and their sensitivities, a nonlinear mathematical programming algorithm is employed to

† The elasticity problem can be written in a general form as

$$\mathbf{K}\mathbf{u} = \mathbf{f} \tag{8.a}$$

where $\mathbf{K}$ is the assembled stiffness matrix, $\mathbf{u}$ is the displacement vector and $\mathbf{f}$ is the applied force vector. Now, if we assume that $w(\mathbf{s})$ is a function of design produce an improved design.

7. *Updating the geometrical design model:* Having the improved values of the design parameters, a modified design is obtained by updating the geometrical model.
8. *Checking the convergence:* If convergence has occurred the process is terminated. Otherwise, the process is repeated with the new parameters of the geometrical model.

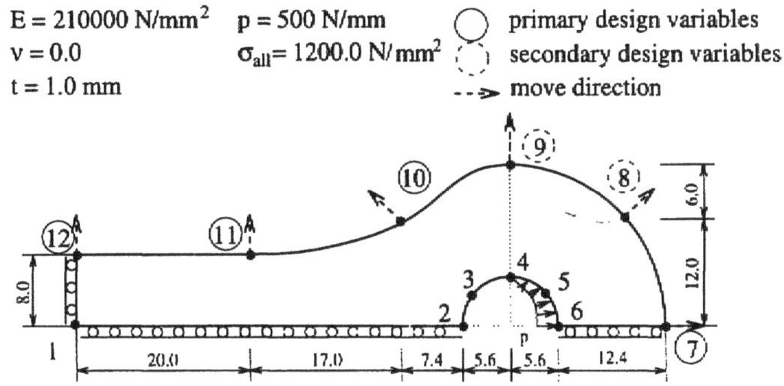

Figure 8.9 *Problem definition of the connecting rod.*

As an example, optimization of a connecting rod is here considered. Geometry, loading, boundary conditions, material properties and the selected key points are given in Figure 8.9. Due to symmetry, only a quarter of the rod is modelled.

→ where **K** is the assembled stiffness matrix, **u** is the displacement vector and **f** is the applied force vector. Now, if we assume that $w(\mathbf{s})$ is a function of design variables (it can be displacement, stress, weight, etc.), the sensitivity w with respect to design parameters $\mathbf{s}$ can be written as

$$\frac{\partial w}{\partial \mathbf{s}} = \frac{\partial w}{\partial \mathbf{u}} \frac{\partial \mathbf{u}}{\partial \mathbf{s}}. \tag{8.b}$$

Differentiating (8.a) with respect to $\mathbf{s}$ and substituting from (8.b), it follows that

$$\frac{\partial w}{\partial \mathbf{s}} = \frac{\partial w}{\partial \mathbf{u}} \left[\mathbf{K}^{-1} \left(\frac{\partial \mathbf{f}}{\partial \mathbf{s}} - \frac{\partial \mathbf{K}}{\partial \mathbf{s}} \mathbf{u} \right) \right]. \tag{8.c}$$

In semi-analytical method $\frac{d\mathbf{f}}{d\mathbf{s}}$ and $\frac{d\mathbf{K}}{d\mathbf{s}}$ are approximated by finite differences $\frac{\Delta \mathbf{f}}{\Delta \mathbf{s}}$ and $\frac{\Delta \mathbf{K}}{\Delta \mathbf{s}}$, respectively. In the global finite difference method the differential of w is computed directly using the expression

$$\frac{\partial w}{\partial \mathbf{s}} \simeq \frac{\Delta w}{\Delta \mathbf{s}} = \frac{w(\mathbf{s} + \Delta \mathbf{s}) - w(\mathbf{s})}{\Delta \mathbf{s}}. \tag{8.d}$$

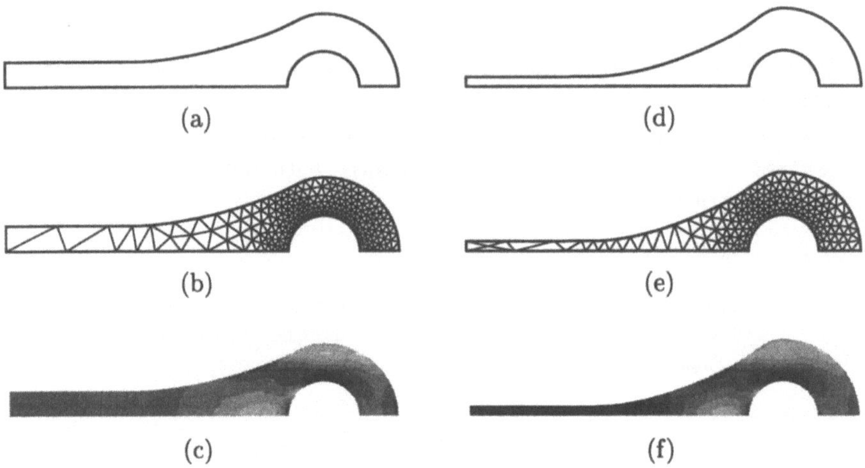

Figure 8.10 (a) Shape of initial design, (b) FE mesh of initial design, (c) stress contours of initial design, (d) shape of optimum design, (e) FE mesh of optimum design and (f) stress contours of optimum design [2]. Courtesy of Dr J. Sienz.

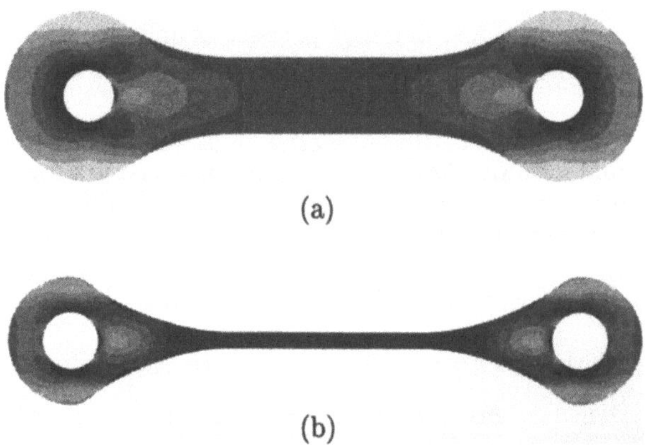

Figure 8.11 Stress contours of the connection rod: (a) Initial design, (b) optimized design [2]. Courtesy of Dr J. Sienz, Dept. of Civil Eng., University of Wales, Swansea, UK.

The objective of the optimization problem is to minimize the volume of material subject to a global stress constraint. The stress constraint is imposed at all Gauss points as well as key points 1, 3, 4, 5, 6 and 12. The initial and optimum shapes, the finite element meshes and the stress contours are illustrated in Figure 8.10. The volume and the maximum stress value of the initial design are $2904mm^3$ and $668.2\,N/mm^2$, respectively. The volume of the optimized shape has reduced to $1072mm^3$ and the maximum stress value of the resulted optimum shape is $1198.9\,N/mm^2$ [2]. In Figure 8.11 the shape and stress contours of the whole rod are illustrated.

8.5.2 Adaptive growth method

The method described here can be considered as an intuitive alternative to shape optimization using the conventional boundary variation method. This method was originally suggested by Mattheck [29, 36]; it is simpler than the boundary variation method as it relies only on conventional analysis tools.

The adaptive growth method tries to simulate the adaptive biological growth of animals and plants in nature. The idea is that 'biological' load carrying systems, as a result of millions of years of evolution have become optimized. This is because of tough competition for energy and living space in nature where only the most efficient biological components have the chance to survive, i.e. those with minimum weight with regard to the required strength.

Amongst the biological structures which are capable of adaptive growth, trees are of special interest. The shape of a tree is a compromise between demanding sufficient light reception (phototropism) which directs the tree towards the light and the tendency to create a large crown on the one hand, and the required mechanical (and biological) supporting members (stem, branches) with a minimum material consumption on other hand.

Mattheck [37] has studied a large number of trees in nature to verify the two mechanical rules that govern the structural growth of trees: the law of 'external load minimization' and the hypothesis of 'constant stress on the surface of trees'. The external load minimization is an attempt to produce a better global stability with regard to the most important loading cases (e.g. tendency to keep the centre of gravity of the crown over the root stock). The constant stress axiom states that biological structures try to grow into a state of constant stress on the surface for different loading at different times. This provides an optimum mechanical state where the danger of failure, which usually starts from a surface, is minimum. In other words, no point on the surface is stressed at a higher level than another. Several examples of constant stress states in biological structures are provided by Mattheck in [36], e.g. tree butts, branch-stem joints, deer antlers, tiger claws, etc.

Based on the axiom of constant stress state the behaviour of trees in wound healing, mechanical contact with foreign bodies (e.g. stones) and welding of living parts of the same species can easily be explained. For example, when a tree is wounded, peak stresses are caused at the border of the wound (notch). Consequently, the tree tries to restore the original state of homogeneous stresses by attaching more material at points of maximum stresses. Similarly, when a mechanical contact load is applied to a tree (e.g. contact with a stone), it reacts by adaptive growth to enlarge the contact surface in order to reduce the localized stress peak.

Failure of a mechanical component, especially due to fatigue loading, is more likely to happen at areas of high local stresses which are often caused by the presence of notches. Therefore, a mechanical design with an homogeneous stress distribution (especially at its surface) is a good one and in some sense optimum. Since in most cases three dimensional stress states exist, the hypothesis of constant stress is often specified by the von Mises stress, as it offers better results in practice [36]. In general the maximum principal stress or strain energy density may also be used.

Several researchers have already suggested algorithms to obtain a homogeneous state of von Mises stresses on the surface of a structure (e.g. Jancu *et al* [38]) and a copy of biological growth (e.g. Umetani *et al* [26] and Huiskes *et al* [39].

Recently, Mattheck [37, 40] suggested a method for computer aided optimization (CAO) of structural shape which mimics the growth of trees†. The main idea behind this method is the controversial idea that not only the biological products are optimum, but the procedures that are followed in nature to achieve them are also 'optimum' procedures, and therefore, can be simulated as optimization mechanisms [36]. However, one may argue that the evolution process in nature is too slow.

In the adaptive growth method proposed by Mattheck and his colleagues the following steps are followed:

1. Create a design proposal of the structure to be optimized and discretize it using finite elements. Obviously, a better proposal reduces the improvement time.
2. Cover the surface of the general finite element mesh with a very thin layer of finite elements. This layer is equivalent to the outermost growth ring in trees and it may be called 'design domain' or 'optimization domain'.
3. Analyse the structure with the given loading, boundary conditions and original material properties‡ using the finite element method and evaluate

---

† It is important to note that trees, unlike bones, do not shrink under reduced stress levels.

‡ In earlier works, to mimic the growth ring, a much smaller Young's modulus for the thin surface layer was considered at this step as well as the next step. See, for example, [29, 36].

the von Mises stresses within the thin layer.

4. Devise a mechanism to allow volumetric swelling of the elements of the surface layer. This can be performed by heating up the surface layer with treating the effective (i.e. von Mises) stresses calculated in the last step, as though they were temperatures. The reference temperature may be considered as zero. The Young's modulus for the surface layer is reduced dramatically at this stage. A reduction factor equal to 1/400 is suggested in [40]. Now, the structure is re-analysed for this fictitious thermal loading while the thermal expansion coefficient of the non-surface layer is taken as zero.
5. The resulting thermal displacements on the surface are added to the dimensions of the structure and the disturbed mesh is corrected in order to restore the same thickness of the thin soft layer as before. One may scale up the thermal displacements by a 'reasonable' large factor to obtain greater efficiency. However, it is noted that increasing the thermal coefficient has the same effect.
6. The steps 3 to 5 are repeated until either one of the functional limitations of the design is violated or all notch stresses disappear and an homogeneous von Mises stress state is obtained.

In Figure 8.12 the algorithm of this method is illustrated.

The adaptive growth approach has been simplified even further by Chen *et al* [41, 42], as they apply the biological growth in all parts of the domain and not only to the thin soft layer of finite elements simulating the growth ring in trees. A similar approach had already been employed by Azegami [43]. Also a shape optimization method has been proposed by Rajan *et al* [44] which is based on fictitious loads acting on an auxiliary structure and using the resulting deformation to update the shape. The simulated biological growth approach suggested by Chen *et al* [42] also makes use of an auxiliary structure with a special boundary condition† according to the restrictions of the shape deformation for design.

Based on the biological growth simulation approaches proposed by Chen *et al* [42], an optimization algorithm can be developed where the possibility of growth as well as atrophy is provided. The objective of this algorithm is mainly stress levelling (i.e. reducing the stress peaks) either in the whole domain or in a given surface layer.

---

† In some problems deciding the appropriate boundary conditions for the growth analysis is not an easy task and this can be seen as one of the main drawbacks of these methods.

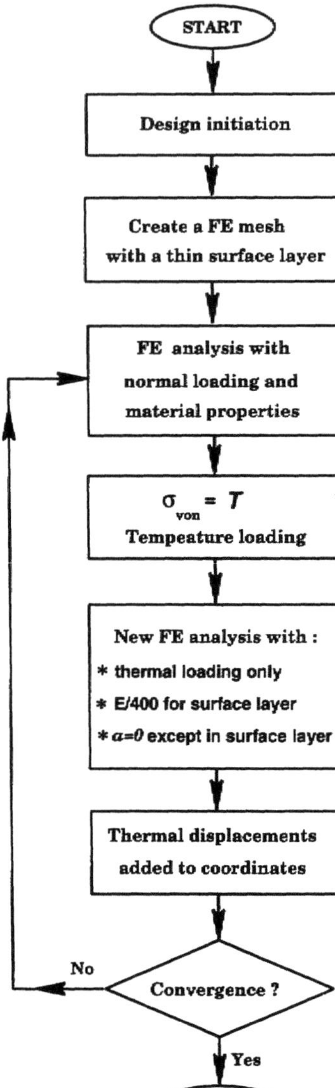

Figure 8.12 *Algorithm of shape optimization using adaptive growth method.*

The optimization problem can be written as

$$\min_{\mathbf{x}\in D} \int_D ||\sigma_{\text{von}}(\mathbf{x}) - \sigma_{\text{bas}}||d\Omega \tag{8.11}$$

where $D \subseteq \Omega$ while D is the design (optimization) domain and Ω denotes the whole domain. $\sigma_{\text{von}}(\mathbf{x})$ is the von Mises stress at a point $\mathbf{x}$ in the design domain D and σ_{bas} is a basic stress used as a reference target. $\|.\|$ can be the Euclidean or the infinity (absolute) norm.

As for the procedure explained earlier, the process comprises two finite element analysis steps. In the first step the effective (von Mises) stresses in the area of interest (the design domain) are evaluated. The second step consists of an incremental growth analysis based on a growth law in which a pseudo-thermal swelling strain is imposed. This swelling strain may be defined as

$$\epsilon_T = \beta\,(\sigma_{\text{von}} - \sigma_{\text{bas}}) \tag{8.12}$$

where β is the incremental growth rate. This swelling strain can be thought of as strain induced by a fictitious thermal loading T such as

$$\epsilon_T = \alpha T \tag{8.13}$$

where α is the thermal expansion coefficient and T is the variation of temperature with respect to a given reference. From (8.12) and (8.13) the temperature T equivalent to the swelling strain has the form

$$T = \eta\,(\sigma_{\text{von}} - \sigma_{\text{bas}}) \tag{8.14}$$

where $\eta = \beta/\alpha$.

At this stage the boundary conditions of the problem are changed according to the geometrical and functional restrictions of the design. One may also use different thermal expansions in different directions (e.g. $\alpha_x = 0$, $\alpha_y = \alpha$).

After carrying out the pseudo-thermal analysis, the incremental displacements are added to the coordinates of the previous step. The flowchart of this algorithm is very similar to the one in Figure 8.12. The only difference is that the design domain is not restricted to the surface layer.

The process is terminated whenever a convergence criteria is satisfied (e.g. when the improvement rate of the objective function is less than a certain value). From a consideration of (8.14) it is realized that the convergence rate is affected by the value of the total incremental growth rate η. On the other hand, we note that ϵ_T in (8.12) and T in (8.14) may have a negative value. Because of this point, contraction (as well as expansion) of the volume (area) of the structure may occur, especially when the design domain is the same as the whole domain. In fact, the value of σ_{bas} is responsible for increasing or decreasing of the volume. That is, by increasing the value of the basic stress σ_{bas}, the volume of the design domain decreases and vice versa.

Because of the dependency of the volume of the structure on the value chosen for σ_{bas} it follows that apart from reducing the local stress peaks (stress levelling), it is also possible to employ this method for the following shape optimization problems: to minimize volume subject to some constraint on the stresses, that is

$$\begin{aligned} &\text{Minimize} && \Omega \\ &\text{subject to} && \sigma \leq \bar{\sigma}, \end{aligned} \tag{8.15}$$

or to minimize the maximum stress value with a given volume of material $\bar{\Omega}$, that is

$$\begin{aligned} &\text{Minimize} && \sigma_{\max} \\ &\text{subject to} && \Omega \leq \bar{\Omega}. \end{aligned} \tag{8.16}$$

It is interesting to note that stresses in (8.15) and (8.16) can be the von Mises stresses at certain points of the domain, the mean von Mises stress† or even a component of stress (e.g. tangential) at a given point.

When our objective is stress levelling without any constraint on the volume, the value of the mean stress can be used as σ_{bas} in each iteration.

For example, to solve the optimization problem (8.15), starting from a feasible design σ_{bas} is gradually increased until the inequality stress constraint is violated. Then by putting the mean stress value of the last iteration in place of σ_{bas} stress levelling may be carried out. This procedure is repeated until convergence has occurred. In the following a couple of examples are presented.

Clamped cantilever beam subject to tip shear loading

The problem definition is given in Figure 8.13. The intensity of the tip shear loading is 6.0. The value of Young's modulus and Poisson's ratio are 2100 and 0.3, respectively.

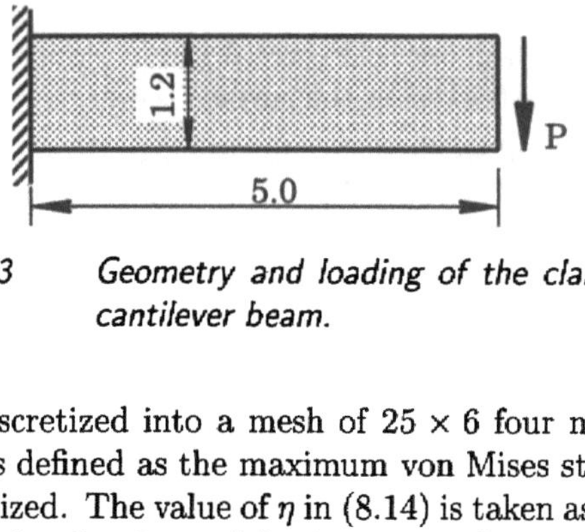

Figure 8.13 *Geometry and loading of the clamped cantilever beam.*

The domain was discretized into a mesh of 25 × 6 four node elements. The objective function is defined as the maximum von Mises stress in the problem domain to be minimized. The value of η in (8.14) is taken as 0.01. The adopted boundary condition for the thermal (or incremental growth) analysis and the initial FE mesh are illustrated in Figure 8.14.

† The mean stress is defined as

$$\sigma_{\text{mean}} = \frac{1}{\Omega}\int_{\Omega} \sigma_{\text{von}}(\mathbf{x})d\Omega.$$

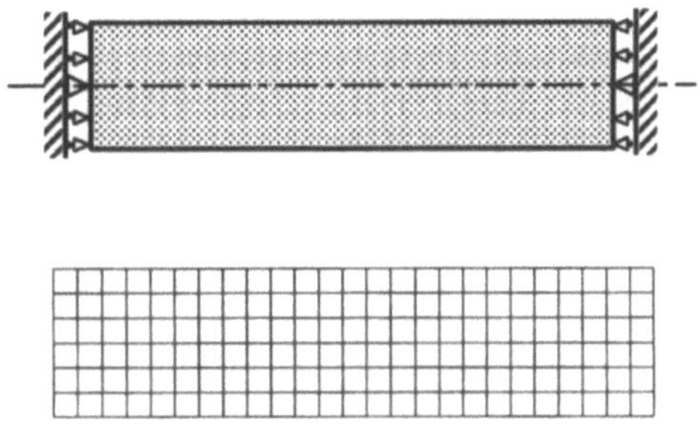

Figure 8.14 *Boundary condition for the thermal analysis and the initial FE mesh.*

The gradual deformation of the finite element mesh and the contours of the equivalent stress ratio to basic stress are shown in Figure 8.15. The iteration history is illustrated in Figure 8.16. A comparison between the values of interest of the initial shape and those of the iterations 50 and 100 are given in Table 8.1.

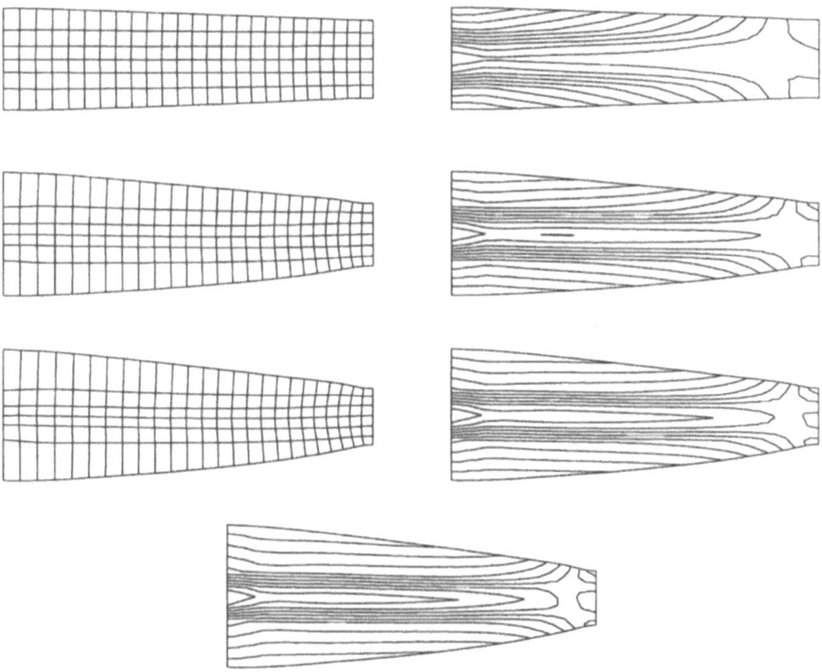

Figure 8.15 *Incremental shape deformation of the FE mesh and contours of the maximum von Mises stress ratio to average stress.*

Design	$\sigma_{\max}$	$\sigma_{\text{av.}}$	$\sigma_{\max}/\sigma_{\text{av.}}$	Area	Strain energy
Initial	82.100	28.444	2.886	6.000	1.764
Iteration 50	42.474	18.908	2.246	6.377	1.069
Iteration 100	27.163	14.159	1.918	7.079	0.719

Table 8.1 *Comparison of values of the objective functions.*

The maximum stress value $\sigma_{\max}$ is the largest von Mises stress calculated at the Gauss points. It is noted that stress distribution in the resulting shape is more even than the initial design. The variation of the objective function during the process is smooth and no oscillations occurred.

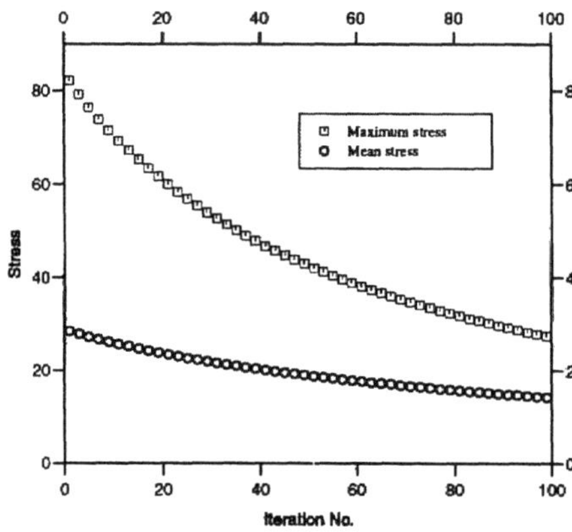

Figure 8.16 *Iteration history of the beam problem.*

Square plate with circular hole

The square plate with a circular hole subject to biaxial in-plane tensile forces is here considered. The loading and the problem definition is shown in Figure 8.17. The intensity of the uniform tensile forces is $p = 1.0$. The Young's modulus and the Poisson's ratio are 2100 and 0.3, respectively. The objective is stress levelling on the whole domain (i.e. minimizing $\sigma_{\max}/\sigma_{\text{av.}}$).

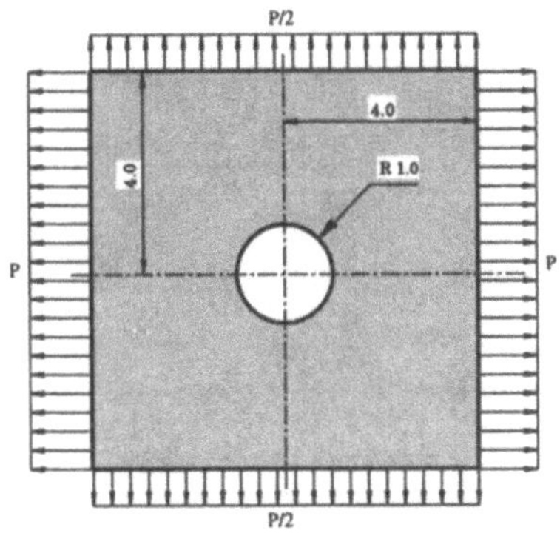

Figure 8.17 Geometry and loading of the square plate with a hole.

By making use of symmetry the problem was solved for a quarter of the plate. In this problem 9-node elements were employed and the domain was discretized using 72 finite elements as shown in the top left picture of Figure 8.20. The boundary conditions for the pseudo-thermal analysis is depicted in Figure 8.18.

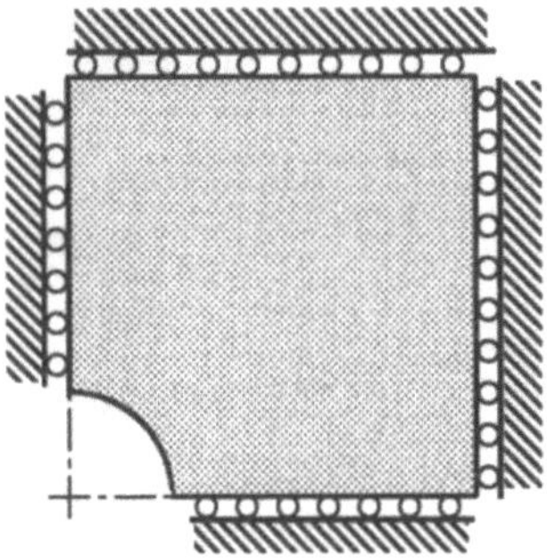

Figure 8.18 Boundary conditions for the thermal analysis.

The history of the maximum and mean stress values is shown in Figure 8.19. The incremental shape deformation of the finite element mesh and the contours of the ratio of the maximum von Mises stress to the mean stress are illustrated in Figures 8.20.

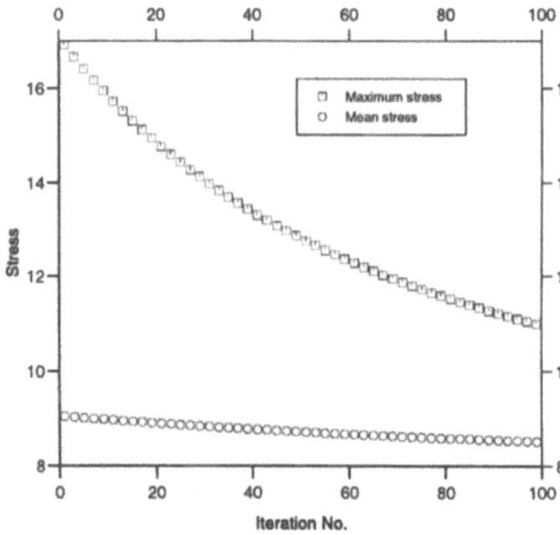

Figure 8.19 *Iteration history.*

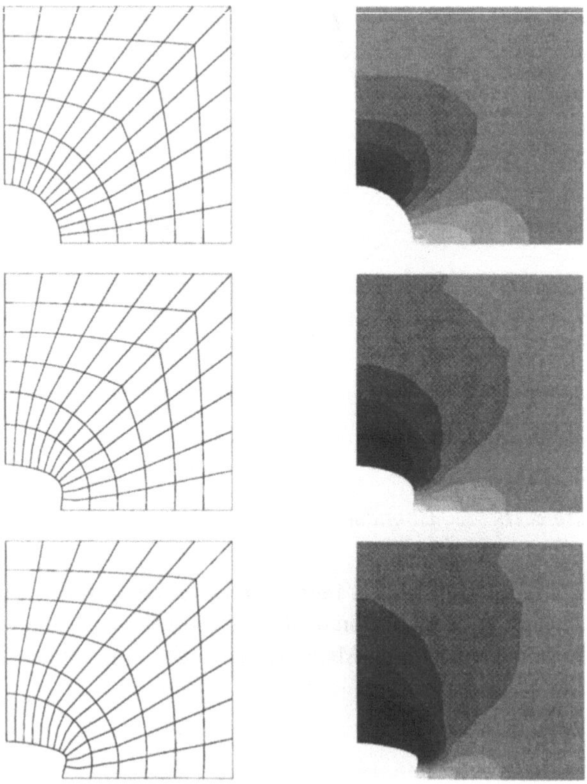

Figure 8.20 *Incremental shape deformation and stress contours.*

Design	σ_{max}	$\sigma_{av.}$	$\sigma_{max}/\sigma_{av.}$	Area	Strain energy
Initial	16.913	9.037	1.871	15.215	0.339
Iteration 50	12.804	8.715	1.469	15.351	0.325
Iteration 100	10.958	8.513	1.287	15.481	0.318

Table 8.2 *Comparison of values of the objective functions for the square plate problem.*

In Table 8.2 the values of interest for the initial shape and those of iterations 50 and 100 are compared. The resulting optimal shape and the stress ratio contours for the whole domain is shown in Figure 8.21. It should be noted that if the objective of this problem was local stress levelling only on the boundary of the hole, then an elliptic shape would have resulted.

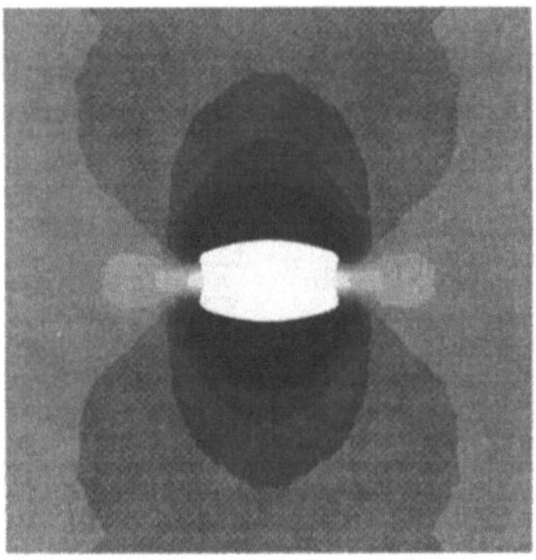

Figure 8.21 *Optimal shape.*

Unlike conventional shape optimization by the boundary variation method, simulated biological methods can be easily extended to three dimensional problems.

8.6 Integrated adaptive topology and shape optimization

The method which is briefly explained here has been suggested by Maute and Ramm [18]. This method is based on an adaptive technique for topology optimization that uses different models for design and analysis. The design model is fixed during the entire optimization process but the analysis model is adapted in each cycle to current material distribution.

Using this method it is easy to carry out an adaptive topology optimization combined with subsequent conventional shape optimization by the boundary variation method. The contours of the material density obtained from the adaptive topology optimization in the analysis model are approximated by Bezier splines. By a threshold method the appropriate contour is then adopted as the boundary of structure. This model can be used either as a new design space for a further topology optimization cycle or as an initial structure for shape optimization. Since the boundaries are defined parametrically, in the shape optimization process the optimal values for the shape parameters are obtained and the result can be mapped onto the design model for further optimization. In this method topology and shape optimization procedures can be carried out sequentially or simultaneously.

The procedure is summarized as follows:

1. Create a fine regular mesh. This is a fixed reference domain called the *design model* which is used for drawing contours of material density and evaluation and control of the area between isolines. Loading, boundary conditions and the initial density of elements are also defined.
2. Create a coarse finite element mesh for the analysis and map the material density values from the design model. This can be done by a simple averaging scheme with respect to the area (volume),

$$\mathbf{s}_j = \frac{\sum_i^n \mathbf{s}_i^0 v_i^0}{\sum_i^n v_i^0} \tag{8.17}$$

   where, $\mathbf{s}_j$ are material parameters of element j in the analysis model (which is called 'active parameters' in [18]) and $\mathbf{s}_i^0$ are the parameters of the design model elements. v_i is the volume (area) of the element i and n is the number of elements which corresponds to the element j of the analysis model. Note that to obtain a reasonable accuracy the design model mesh has to be reasonably fine. In general the parameters of the analysis elements $\mathbf{S}$ and the design elements $\mathbf{S}^0$ are related to each other with a linking matrix, that is

$$\mathbf{S} = \mathbf{L}\mathbf{S}^0 \tag{8.18}$$

3. Carry out the topology optimization process and find the material density of elements in the analysis model.
4. Remap the resulting density values from the analysis model to the design model (background mesh). This can be done by interpolation using finite element shape functions.
5. Plot the polygon contours (isolines) of density by linear interpolation of the density values at the nodal points of the design mesh.
6. Substitute the contour polygons with suitable smooth curves using splines or Lagrangian curves. The cubic, Bezier or B-spline parameterization of

the contours provides the possibility of using the resulting geometry in a conventional shape optimization procedure. If $\mathbf{p}_i$ denote the points of the contours and $\mathbf{q}_i$ denote the corresponding points of the Bezier spline, the coordinates of the control points can be obtained via a minimization problem:

$$\text{Minimize} \quad \sum_{i=1}^{n} ||\mathbf{p}_i - \mathbf{q}_i|| \tag{8.19}$$

where $||.||$ denotes the Euclidean norm and n is the number of vertices of a given isoline polygon. $\mathbf{q}_i$ can be expressed in terms of control points $\mathbf{b}_j$, $(j = 1, \cdots, 4)$ and blending functions ϕ_j:

$$\mathbf{q}_i = \sum_j \phi_j(h_i)\mathbf{b}_j, \tag{8.20}$$

where the local coordinate h_i of each point i is calculated by the expression

$$h_i = \ell_i/\ell_n \tag{8.21}$$

and

$$\ell_i = \sum_{k=1}^{i} ||\mathbf{p}_{k-1} - \mathbf{p}_k||. \tag{8.22}$$

Differentiating (8.19) with respect to the coordinates of the control points of the Bezier spline results in a system of linear equations which can be solved easily.

The approximation algorithm begins with only one spline and the optimum location of the control points are determined. Then the approximation error is calculated by the

$$\text{err} = \frac{1}{\ell_n}\sum_{i=1}^{n} ||\mathbf{p}_i - \mathbf{q}_i||. \tag{8.23}$$

If the evaluated error is larger than a given limit, then the number of spline control points is increased and the procedure is repeated. For more details the reader is directed to consult references [18] and [45].

7. Areas with density values below a certain threshold level are considered as voids and neglected. Also, the hole areas below a certain size can be neglected. It is interesting to note that since the original optimization variables are kept in the design model, in this method a hole neglected in one stage does not mean that it is neglected in all following cycles and it is possible that material be accumulated again in the already existing holes.

8. An adaptive mesh refinement analogous to the methods of h-adaptivity used in the finite element method to reduce the error in the energy norm is used. The objective of mesh refinement is to decrease the number of finite elements in the analysis model and consequently to decrease the number of

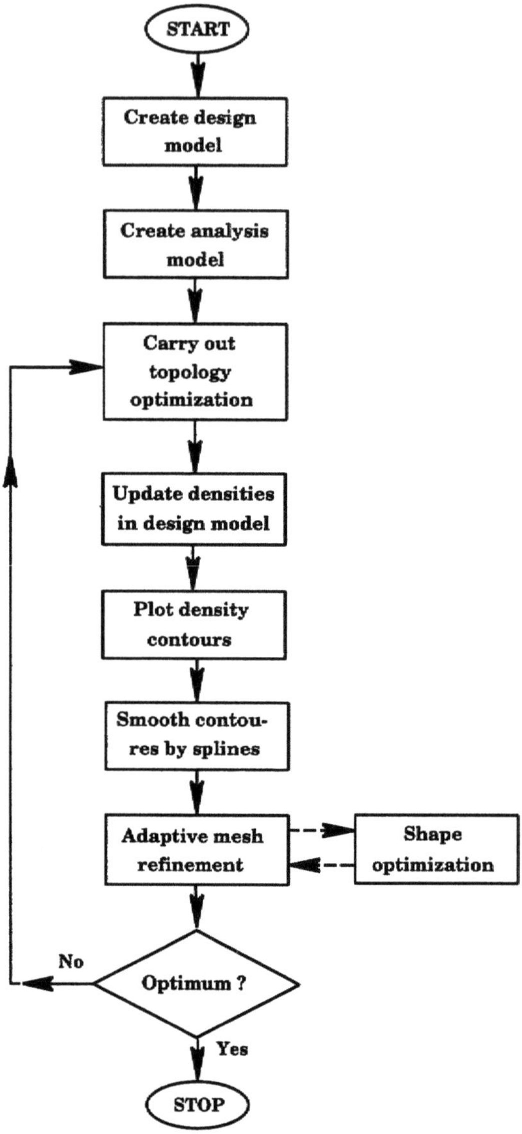

Figure 8.22 *Algorithm for adaptive topology and shape optimization.*

design variables. In addition the size and orientation of the elements of the analysis model are adapted to the material distribution of the design model. The local refinement of the finite element mesh is controlled by the material density distribution in the design model. Assuming that the areas of the patches in the design model are of equal size, the local refining indicator η

is defined with

$$\eta = \begin{cases} \eta_{\max} - (1-\alpha)^2 (\eta_{\max} - \eta_{\min}) & \text{for } \rho < \rho_{\lim} \\ \eta_{\max} & \text{for } \rho \geq \rho_{\lim} \end{cases} \tag{8.24}$$

where $\eta_{\max}$ and $\eta_{\min}$ are upper and lower bounds of the refining indicator η and $\alpha = \rho/\rho_{\lim}$ is the relative density. $\rho_{\lim}$ is the limiting value of the material density. For the density values above this value the refining factor is set to the upper bound $\eta_{\max}$. From (8.24) it is observed that for areas with a very low material density ($\rho \simeq 0$), the refinement indicator is equal to $\eta_{\min}$ and a coarse finite element mesh is generated. However, for the dense areas with $\rho \simeq \rho^0$ a fine finite element mesh with the refinement indicator equal to $\eta_{\max}$ is generated.

Regarding the distribution of the refining indicator, a new quadrilateral mesh is generated by a fully automatic mesh generator based on the advancing front method [2, 46]. During the adaptive optimization process the maximum refining indicator $\eta_{\max}$ is gradually increased until the size of the smallest finite elements of the analysis model is equal to the size of the regular patches (elements) of the design model.

9. The generated mesh in the last step, may be used for a conventional shape optimization by the boundary variation technique. This improves the numerical efficiency and the quality of the results of the optimization process. To obtain further numerical efficiency, one can make use of adaptivity in the geometrical shape definition. This can be achieved by gradually decreasing the acceptable error in the spline approximation. Consequently, the number of control points defining the geometry of the boundaries of the structure are gradually increased. This method may be called 'general adaptive shape optimization' [45].

   It is possible to start the process with shape optimization followed by adaptive topology and shape optimization cycles. In this case the result of an shape optimization step is mapped onto the design model just as the result of a topology optimization step. We note that in the 'adaptive topology approach' the shape optimization step does not exist.

   The shape optimization problem can be solved by mathematical programming methods (e.g. SQP), where the design variables of the optimization problem are the coordinates of the control points and the assumed volume is a constraint.

10. The steps 3 to 9 are repeated until the optimum layout is obtained. The convergence criteria may be constructed based on the difference between the material densities of the elements of the fixed design model in two successive iterations. It might also be constructed based on the difference between the values of the objective functions in successive iterations. The algorithm of the adaptive topology (and shape) optimization is illustrated in Figure 8.22.

Figure 8.23 illustrates this method when applied to optimize a cantilever beam subject to a point load on the bottom left corner [45].

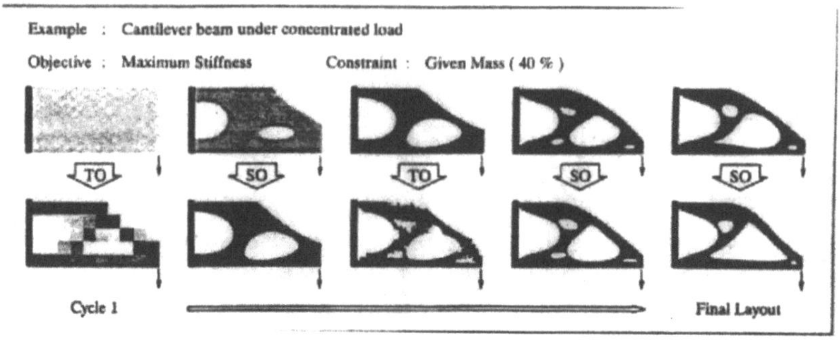

Figure 8.23 *General shape optimization by adaptive topology optimization and boundary variation [45]. Courtesy of Dr K. Maute, University of Stuttgart, Germany.*

The topology and shape optimization may be carried out in only one analysis model, where shape and material parameters are considered as the optimization variables.

As an example of integrated topology and shape optimization, Example 2 of Chapter 4 is again considered here. The image processing of the image obtained from the topology optimization module was carried out as illustrated in Figure 8.4. The layout of Figure 8.4(c) was considered and the boundary points were selected manually and the boundary was smoothed using B-splines. Then, the domain of problem is discretized into a mesh of 9-node finite elements. By making use of symmetry the problem was solved for half of the domain. The finite element mesh is given in Figure 8.24.

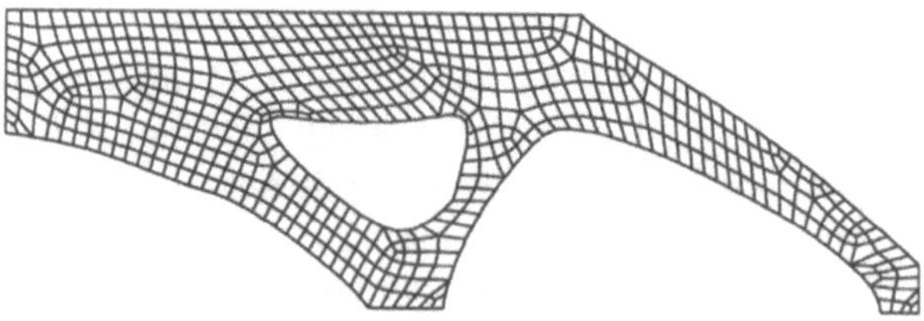

Figure 8.24 *Finite element mesh.*

A shape optimization process with the objective of stress levelling was carried out. For this purpose the adaptive growth method was utilized. The initial and final shapes as well as the stress contours are illustrated in Figures 8.25 and 8.26, respectively.

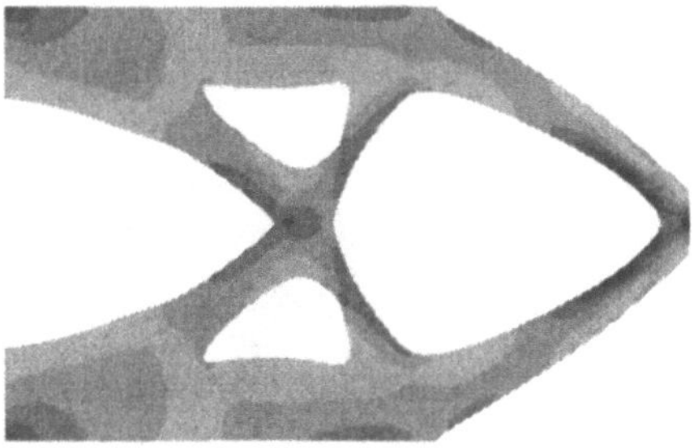

Figure 8.25 *Initial shape and stress contours.*

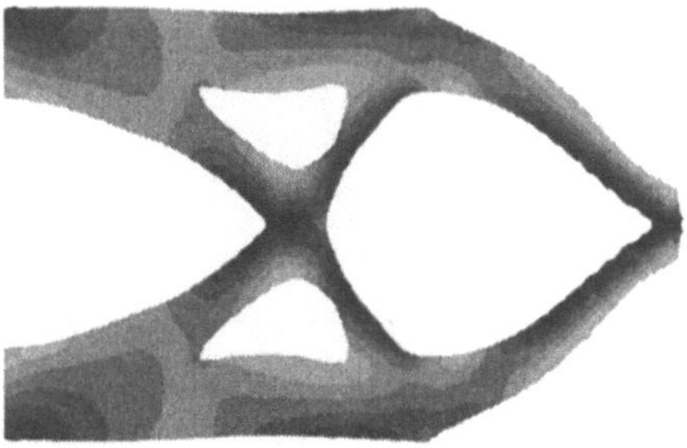

Figure 8.26 *Stress contours of the resulting optimal shape.*

It is observed that the stresses in the final shape are distributed more evenly than the initial shape. In Table 8.3 the values of interest are compared.

Design	$\sigma_{\max}$	$\sigma_{\mathrm{av.}}$	$\sigma_{\max}/\sigma_{\mathrm{av.}}$	Area	Strain energy
Initial	9386.1	1582.4	5.931	35.92	572.6
Iteration 10	7532.5	1560.6	4.827	35.95	552.3
Iteration 20	6050.9	1537.9	3.935	36.00	535.0
Iteration 30	5031.5	1520.2	3.310	36.06	520.4
Iteration 40	4412.0	1507.7	2.926	36.13	507.7

Table 8.3 *Comparison of values.*

8.7 Final thoughts

Just before finishing this text, we would like to share a few final thoughts. Structural topology optimization is still a rich area of research with new work being completed and published all the time. Interested readers should follow the excellent set of conferences of ISSMO (International Society for Structural and Multidisciplinary Optimization) [47, 48, 49]. We would also like to draw attention to two recent PhD theses from Beckers and Maute which contain much interesting material which unfortunately appeared too late to include in this book [50, 51]. Finally, apart from the existing journals: *Structural Optimization* and *Engineering Optimization*, there is a new international journal entitled *Design Optimization: International Journal for Product and Process Optimization* which will focus on practical applications of optimization — this will be of interest to readers of this text.

References

[1] Tworzydlo W.W. and Oden J.T., Towards an automated environment in computational mechanics, *Comp. Meth. Appl. Mech. Engng*, **104**, 87–143 (1993)

[2] Sienz J., *Integrated structural modelling, adaptive analysis and shape optimization*, C/ph/181/94, Dept. of Civil Engineering, University College of Swansea, Swansea, UK (1994)

[3] Bendsøe M.P., *Optimization of structural topology, shape, and material.* Springer, Berlin (1995)

[4] Bendsøe M.P. and Rodrigues H.C., Integrated topology and boundary shape optimization of 2-D solids, *Computer Methhods in Applied Mchanics and Engineering*, **87**, 15–34 (1991)

[5] Bremicker M., Chirehdast M., Kikuchi N. and Papalambros P.Y., Integrated topology and shape optimization in structural design, *Mech. Struct. Mech.*, **19**(4), 551–587 (1991)

[6] Ballard D. and Brown C., *Computer vision.* Printice Hall, Englewood Cliffs, New Jersey (1982)

[7] Rozvany G.I.N. (ed.), *Shape and layout optimization of structural systems and optimality criteria methods.* Springer-Verlag, CISM, Udine (1992)

[8] Rozvany G.I.N. and Zhou M., Layout and generalized shape optimization by iterative COC methods, in *Optimization of large structural systems*, edited by Rozvany G.I.N., vol. 1, pp. 103–120, Berchtesgaden (1993). NATO DFG ASI

[9] Rozvany G.I.N. and Zhou M., Optimality criteria methods for large discretized systems, in *Advaces in design optimization*, edited by Adeli H., pp. 41–108. Chapman & Hall, London (1994)

[10] Rozvany G.I.N., Bendsøe M.P. and Kirsch U., Layout optimization of structures, *Appl. Mech. Rev.*, **48**(2), 41–119 (1995)

[11] Eschenauer H.A., Schumacher A. and Vietor T., Decision makings for initial designs made of advanced materials, in *Topology design of structures*, edited by Bendsøe M.P. and Mota Soares C.A., pp. 469–480, Dordrecht NL (1993). NATO/ARW, Kluwer Academic Publishers

[12] Eschenauer H.A., Kobelev V.V. and Schumacher A., Bubble method for topology and shape optimization of structures, *Structural Optimization*, **8**, 42–51 (1994)

[13] Zhou M. and Rozvany G.I.N., DCOC: an optimality criteria method for large systems. Part I: theory, *Structural Optimization*, **5**, 12–25 (1992)

[14] Rozvany G.I.N., Zhou M. and Sigmund O., Optimization of topology, in *Advaces in design optimization*, edited by Adeli H., pp. 340–399. Chapman & Hall, London (1994)

[15] Zhou M. and Haftka R.T., A comparison of optimality criteria methods for stress and displacement constraints, *Computer meth. in appl. mech. and engng.*, **124**, 253–271 (1995)

[16] Farin G., *Curves and surfaces for computer aided geometric design.* Academic Press, London (1990)

[17] Olhoff N., Bendsøe M. and Rasmussen J., On CAD-integrated structural topology and design optimization, *Comp. Meth. in Appl. Mech.*, **89**, 259–279 (1991)

[18] Maute K. and Ramm E., Adaptive topology optimization, *Structural Opimization*, **10**, 100–112 (1995)

[19] Chirehdast M., Gea H.C., Kikuchi N. and Papalambros P.Y., Structural Configuration examples of an integrated optimal design process, *J. of Mechanical Design*, **1**, 997–1004 (1994)

[20] Sigmund O., *Design of material structures using topology optimization*, Ph.D. thesis, Tech. Univ. of Denmark (1994)

[21] Jog C.S., *Variable-topology shape optimization of linear elastic structures*, Ph.D. thesis, Dept. of Theoretical and Applied Mechanics, Univ. of Illinois at Urbana-Champaign, USA (1993)

[22] Pratt W.K., *Digital signal processing.* John Wiley & Sons, New York (1991)

[23] Jog C.S., Haber R.B. and Bendsøe M.P., Variable topology-shape optimization with a constraint on the perimeter, in *20'th ASME Design Automatio Conference*, Minneapolis (1994)

[24] Haftka R.T. and Grandhi R.V., Structural shape optimization - a survey, *Comp. Meth. Appl. Mech. Engrg.*, **57**, 91–106 (1986)

[25] Eizadian D. and Trompette A.J., Shape optimization of bidimensional structures by the boundary element method, in *CAD/CAM, Robotics and Automation conference*, University of Arizona, Tucson, AZ (1985)

[26] Umetani Y. and Hirai S., Shape optimization of beams subject to displacement restrictions on the basis of the growing-reforming procedure, *Bull. JSME*, **21**, 1113–1119 (1978)

[27] Hemada M., Seguchi Y. and Tada Y., Shape determination problems of structures by the inverse variational principle, *Bull. JSME*, **10**, 23–184 (1980)

[28] Schnack E., Computer simulation of an experimental method for notch shape optimization, in *Simulation in engineering science*, edited by Burger J. and Jarny Y., pp. 311–316. Elsevier Science Publishers, Amsterdam (1983)

[29] Mattheck C. and Burkhardt S., A new method of structural shape optimization based on biological growth, Tech. rep., Institute für Material - Kernforschungszentrum Karlsruhe GmbH (1989)

[30] NAG, The NAG fortran library manual, Tech. rep., Numerical Algorithms Group Ltd, Oxford, UK (1988)

[31] Vanderplaats G.N., *DOT, User manual*, VMA Engineering, Colorado Springs, Colorado, USA (1994)

[32] Svanberg K., The method of moving asymptotes - a new method for structural optimization, *Int. J. Num. Meth. Engng.*, **23**, 359–373 (1987)

[33] Sienz J., Hassani B. and Hinton E., Structural optimization, in *Advanced finite element solution techniques in innovative computer architectures*, edited by Bugeda G. and Papadrakakis M. CIMNE Barcelona (1996), (To appear in)

[34] Zienkiewicz O.C. and Zhu J.Z., A simple error estimator and adaptive procedure for practical engineering analysis, *Int. J. Num. Meth. Engng.*, **24**, 337–357 (1987)

[35] Zienkiewicz O.C. and Zhu J.Z., Superconvergent derivative recovery techniques and a posteriori error estimates: parts 1 and 2, *Int. J. Num. Meth. Engng.*, **33**, 1331–1382 (1992)

[**36**] Mattheck C., Engineering components grow like trees, *Mat.-wiss. u. Werkstofftech*, **21**, 143–168 (1990)

[**37**] Mattheck C., *Trees – The mechanical design*. Springer Verlag, Heidelberg, New York (1991)

[**38**] Jancu G. and Schnack E., Control of the Von Mises stress with dynamic programming, in *GAMM-Meeting*, Siegen (1988)

[**39**] Huiskes R., Weinans H., Grootenboer H.J., Dalstra M., Fudala B. and Slooff T.J., Adaptive bone-remodeling theory applied to prosthetic-design analysis, *J. Biomech.*, **22**, 1135–1150 (1987)

[**40**] Walther F. and Mattheck C., Local stiffening and sustaining of shell and plate structures by SKO and CAO, *Structural Optimization*, pp. 181–188 (1993)

[**41**] Chen J.L. and Tsai W.C., A study of simulated biological growth approaches for shape optimization, in *Computational methods in engineering advances and applications*, pp. 1174–1179. World Scientific, Singapore (1992)

[**42**] Chen J.L. and Tsai W.C., Shape optimization by using simulated biological growth approaches, *AIAA Journal*, **31**(11), 2143–2147 (1993)

[**43**] Azegami H., A proposal of a shape-optimization method using a constitutive equation of growth, *JSME Int. J.*, **33**(1), 64–71 (1990)

[**44**] Rajan S.D. and Belegundu A.D., Shape optimal design using fictitious loads, *AIAA Journal*, **27**(1), 102–107 (1989)

[**45**] Maute K. and Ramm E., General shape optimization - an integrated model for topology and shape optimization, in *First world congress of Structural and Multidisciplinary Optimization*, edited by Rozvany G.I.N., Goslar, Lower Saxony,Germany (1995)

[**46**] Peraire J., Vahdati M., Morgan K. and Zienkiewicz O.C., Adaptive remeshing for compressible flow computations, *J. Comp. Phys.*, **72**, 449–466 (1987)

[**47**] Olhoff N. and Rozvany G.I.N., *Proceedings of the First World Congress of Structural and Multidisciplinary Optimization*, vol. 1. Pergamon, Goslar, Germany (1995)

[**48**] Gutkowski W. and Mroz Z., *Proceedings of the Second World Congress of Structural and Multidisciplinary Optimization*. Zakopane, Poland (1997)

[**49**] Steven G., Querin O.M., Guan H. and Xie Y.M., *Proceedings of The Australasian Conference on Structural Optimization*. Oxbridge Press, Sydney Australia (1997)

[50] Beckers M., *Optimisation de structures en variables discrètes*, Ph.D. thesis, Universite de Liege, Faculte de Sciences (1997)

[51] Maute K., *Topologie— und Formoptimierung von dünnwandigen Tragwerken*, Ph.D. thesis, Universität Stuttgart, Institut für Baustatik (1998)

Appendix A

Using tensor notation the constitutive law of linear elasticity, in general can be written as

$$\sigma_{ij} = E_{ijkl}\epsilon_{kl} \tag{A.1}$$

where in the case of general orthotropicity (i.e. symmetry with respect to three mutually orthogonal planes) the elasticity tensor E_{ijkl} is reduced to

$$E_{ijkl} = \begin{bmatrix} E_{1111} & E_{1122} & E_{1133} & 0 & 0 & 0 \\ E_{1122} & E_{2222} & E_{2233} & 0 & 0 & 0 \\ E_{1133} & E_{2233} & E_{3333} & 0 & 0 & 0 \\ 0 & 0 & 0 & E_{1212} & 0 & 0 \\ 0 & 0 & 0 & 0 & E_{2323} & 0 \\ 0 & 0 & 0 & 0 & 0 & E_{3131} \end{bmatrix}. \tag{A.2}$$

It is sometimes convenient to introduce a concentrated notation and to rewrite (A.1) in the compact matrix form

$$\sigma_\alpha = E_{\alpha\beta}\epsilon_\beta \qquad (\alpha, \beta = 1, 2, \cdots, 6) \tag{A.3}$$

where

$$\sigma_1 = \sigma_{11}, \quad \sigma_2 = \sigma_{22}, \quad \sigma_3 = \sigma_{33}, \quad \sigma_4 = \sigma_{23}, \quad \sigma_5 = \sigma_{13}, \quad \sigma_6 = \sigma_{12},$$

$$\epsilon_1 = \epsilon_{11}, \quad \epsilon_2 = \epsilon_{22}, \quad \epsilon_3 = \epsilon_{33}, \quad \epsilon_4 = \epsilon_{23}, \quad \epsilon_5 = \epsilon_{13}, \quad \epsilon_6 = \epsilon_{12}$$

and the two-index components $D_{\alpha\beta}$ are obtained from the components of E_{ijkl} by the following algorithm:

$$11 \to 1, \quad 22 \to 2, \quad 33 \to 3, \quad 23 \to 4, \quad 13 \to 5, \quad 12 \to 6.$$

So, using the above notation, the matrix of elastic moduli (A.2) can be written as

$$\mathbf{D} = \begin{bmatrix} D_{11} & D_{12} & D_{13} & 0 & 0 & 0 \\ D_{12} & D_{22} & D_{23} & 0 & 0 & 0 \\ D_{13} & D_{23} & D_{33} & 0 & 0 & 0 \\ 0 & 0 & 0 & D_{44} & 0 & 0 \\ 0 & 0 & 0 & 0 & D_{55} & 0 \\ 0 & 0 & 0 & 0 & 0 & D_{66} \end{bmatrix}. \tag{A.4}$$

In the case of two dimensional problem (A.4) is reduced to

$$\mathbf{D} = \begin{bmatrix} D_{11} & D_{12} & 0 \\ D_{12} & D_{22} & 0 \\ 0 & 0 & D_{66} \end{bmatrix}. \tag{A.5}$$

Appendix B

The results obtained by the finite element solution of the homogenization equation for square unit cells with rectangular holes are given in Table B.1. The material of the solid part of the cell is assumed to be isotropic with Young's modulus $E = 0.91$ and the Poisson's ratio $\nu = 0.3$.

a	b	D_{11}^H	D_{12}^H	D_{22}^H	D_{66}^H	a	b	D_{11}^H	D_{12}^H	D_{22}^H	D_{66}^H
0.0	0.0	1.0000	0.3000	1.0000	0.3500	0.3	0.0	0.9881	0.2603	0.8676	0.3323
0.0	0.1	0.9850	0.2955	0.9987	0.3482	0.3	0.1	0.9385	0.2493	0.8258	0.2990
0.0	0.2	0.9386	0.2816	0.9945	0.3423	0.3	0.2	0.8591	0.2294	0.7941	0.2661
0.0	0.3	0.8676	0.2603	0.9881	0.3323	0.3	0.3	0.7644	0.2039	0.7644	0.2313
0.0	0.4	0.7802	0.2341	0.9802	0.3183	0.3	0.4	0.6637	0.1756	0.7367	0.1956
0.0	0.5	0.6840	0.2052	0.9716	0.3004	0.3	0.5	0.5620	0.1467	0.7115	0.1597
0.0	0.6	0.5853	0.1756	0.9627	0.2789	0.3	0.6	0.4617	0.1179	0.6896	0.1237
0.0	0.7	0.4866	0.1460	0.9538	0.2536	0.3	0.7	0.3628	0.0899	0.6714	0.0870
0.0	0.8	0.3891	0.1167	0.9450	0.2240	0.3	0.8	0.2638	0.0628	0.6572	0.0493
0.0	0.9	0.2879	0.0863	0.9359	0.1866	0.3	0.9	0.1580	0.0357	0.6464	0.0140
0.0	1.0	0.0000	0.0000	0.9100	0.0000	0.3	1.0	0.0000	0.0000	0.6370	0.0000
0.1	0.0	0.9987	0.2955	0.9850	0.3482	0.4	0.0	0.9802	0.2341	0.7802	0.3183
0.1	0.1	0.9689	0.2882	0.9689	0.3360	0.4	0.1	0.9218	0.2222	0.7319	0.2745
0.1	0.2	0.9078	0.2722	0.9544	0.3196	0.4	0.2	0.8352	0.2013	0.6961	0.2348
0.1	0.3	0.8258	0.2493	0.9385	0.2990	0.4	0.3	0.7367	0.1756	0.6637	0.1956
0.1	0.4	0.7319	0.2222	0.9218	0.2745	0.4	0.4	0.6348	0.1487	0.6348	0.1582
0.1	0.5	0.6332	0.1928	0.9046	0.2466	0.4	0.5	0.5335	0.1222	0.6100	0.1231
0.1	0.6	0.5336	0.1625	0.8874	0.2153	0.4	0.6	0.4342	0.0968	0.5896	0.0905
0.1	0.7	0.4350	0.1317	0.8706	0.1797	0.4	0.7	0.3367	0.0726	0.5736	0.0596
0.1	0.8	0.3357	0.0999	0.8543	0.1364	0.4	0.8	0.2398	0.0496	0.5615	0.0307
0.1	0.9	0.2261	0.0648	0.8390	0.0749	0.4	0.9	0.1381	0.0271	0.5527	0.0074
0.1	1.0	0.0000	0.0000	0.8190	0.0000	0.4	1.0	0.0000	0.0000	0.5460	0.0000
0.2	0.0	0.9945	0.2816	0.9386	0.3423	0.5	0.0	0.9716	0.2052	0.6840	0.3004
0.2	0.1	0.9544	0.2722	0.9078	0.3196	0.5	0.1	0.9046	0.1928	0.6332	0.2466
0.2	0.2	0.8833	0.2540	0.8833	0.2947	0.5	0.2	0.8122	0.1714	0.5953	0.2013
0.2	0.3	0.7941	0.2294	0.8591	0.2661	0.5	0.3	0.7115	0.1467	0.5620	0.1597
0.2	0.4	0.6961	0.2013	0.8352	0.2348	0.5	0.4	0.6100	0.1222	0.5335	0.1231
0.2	0.5	0.5953	0.1714	0.8122	0.2013	0.5	0.5	0.5101	0.0992	0.5101	0.0917
0.2	0.6	0.4949	0.1409	0.7907	0.1654	0.5	0.6	0.4124	0.0778	0.4917	0.0646
0.2	0.7	0.3955	0.1102	0.7712	0.1263	0.5	0.7	0.3164	0.0578	0.4776	0.0407
0.2	0.8	0.2951	0.0794	0.7655	0.0818	0.5	0.8	0.2213	0.0388	0.4671	0.0197
0.2	0.9	0.1857	0.0476	0.7416	0.0303	0.5	0.9	0.1232	0.0204	0.4597	0.0043
0.2	1.0	0.0000	0.0000	0.7280	0.0000	0.5	1.0	0.0000	0.0000	0.4550	0.0000

Table B.1 *Elements of homogenized constitutive matrix for cellular material with square microcells and rectangular voids*

a	b	D^H_{11}	D^H_{12}	D^H_{22}	D^H_{66}
0.6	0.0	0.9627	0.1756	0.5853	0.2789
0.6	0.1	0.8874	0.1625	0.5336	0.2153
0.6	0.2	0.7907	0.1409	0.4949	0.1654
0.6	0.3	0.6896	0.1179	0.4617	0.1237
0.6	0.4	0.5896	0.0968	0.4342	0.0905
0.6	0.5	0.4917	0.0778	0.4124	0.0646
0.6	0.6	0.3955	0.0606	0.3955	0.0441
0.6	0.7	0.3008	0.0447	0.3828	0.0272
0.6	0.8	0.2070	0.0296	0.3736	0.0129
0.6	0.9	0.1119	0.0151	0.3674	0.0027
0.6	1.0	0.0000	0.0000	0.3640	0.0000
0.7	0.0	0.9538	0.1460	0.4866	0.2536
0.7	0.1	0.8706	0.1317	0.4350	0.1797
0.7	0.2	0.7712	0.1102	0.3955	0.1263
0.7	0.3	0.6714	0.0899	0.3628	0.0870
0.7	0.4	0.5736	0.0726	0.3367	0.0596
0.7	0.5	0.4776	0.0578	0.3164	0.0407
0.7	0.6	0.3828	0.0447	0.3008	0.0272
0.7	0.7	0.2891	0.0328	0.2891	0.0168
0.7	0.8	0.1962	0.0215	0.2808	0.0082
0.7	0.9	0.1031	0.0107	0.2755	0.0018
0.7	1.0	0.0000	0.0000	0.2730	0.0000
0.8	0.0	0.9450	0.1167	0.3891	0.2240
0.8	0.1	0.8543	0.0999	0.3357	0.1364
0.8	0.2	0.7655	0.0794	0.2951	0.0818
0.8	0.3	0.6572	0.0628	0.2638	0.0493
0.8	0.4	0.5615	0.0496	0.2398	0.0307
0.8	0.5	0.4671	0.0388	0.2213	0.0197
0.8	0.6	0.3736	0.0296	0.2070	0.0129
0.8	0.7	0.2808	0.0215	0.1962	0.0082
0.8	0.8	0.1886	0.0141	0.1886	0.0045
0.8	0.9	0.0967	0.0069	0.1839	0.0011
0.8	1.0	0.0000	0.0000	0.1820	0.0000
0.9	0.0	0.9359	0.0863	0.2879	0.1866
0.9	0.1	0.8390	0.0648	0.2261	0.0749
0.9	0.2	0.7416	0.0476	0.1857	0.0303
0.9	0.3	0.6464	0.0357	0.1580	0.0140
0.9	0.4	0.5527	0.0271	0.1381	0.0074
0.9	0.5	0.4597	0.0204	0.1232	0.0043
0.9	0.6	0.3674	0.0151	0.1119	0.0027
0.9	0.7	0.2755	0.0107	0.1031	0.0018
0.9	0.8	0.1839	0.0069	0.0967	0.0011
0.9	0.9	0.0925	0.0035	0.0925	0.0005
0.9	1.0	0.0000	0.0000	0.0910	0.0000
1.0	0.0	0.9100	0.0000	0.0000	0.0000
1.0	0.1	0.8190	0.0000	0.0000	0.0000
1.0	0.2	0.7280	0.0000	0.0000	0.0000
1.0	0.3	0.6370	0.0000	0.0000	0.0000
1.0	0.4	0.5460	0.0000	0.0000	0.0000
1.0	0.5	0.4550	0.0000	0.0000	0.0000
1.0	0.6	0.3640	0.0000	0.0000	0.0000
1.0	0.7	0.2730	0.0000	0.0000	0.0000
1.0	0.8	0.1820	0.0000	0.0000	0.0000
1.0	0.9	0.0910	0.0000	0.0000	0.0000
1.0	1.0	0.0000	0.0000	0.0000	0.0000

Table B.1 *Elements of homogenized constitutive matrix for cellular material with square microcells and rectangular voids (continued)*

Appendix C

The following polynomials are obtained from least squares smoothing:

C.1 Bi-cubic complete polynomial

$$D_{11}^{H}(a,b) = 1.03470 - 0.19507a - 0.44815b + 0.17949a^2 - 0.83233ab - 0.46682b^2 - 0.08561a^3 + 0.19546a^2b + 0.68235ab^2 - 0.04131b^3 \quad (C.1)$$

$$D_{12}^{H}(a,b) = 0.31345 - 0.12080a - 0.12080b - 0.17639a^2 - 0.26954ab - 0.17639b^2 \; 0.00835a^3 + 0.27454a^2b + 0.27454ab^2 + 0.00835b^3 \quad (C.2)$$

$$D_{22}^{H}(a,b) = 1.03470 - 0.44815a - 0.19507b - 0.46682a^2 - 0.83233ab + 0.17949b^2 - 0.04131a^3 + 0.68235a^2b + 0.19546ab^2 - 0.08561b^3 \quad (C.3)$$

$$D_{66}^{H}(a,b) = 0.37940 - 0.17684a - 0.17684b + 0.12103a^2 - 1.03828ab + 0.12103b^2 - 0.24676a^3 + 0.65449a^2b + 0.65449ab^2 - 0.24676b^3 \quad (C.4)$$

C.2 Bi-quartic complete polynomial

$$D_{11}^{H}(a,b) = 1.00382 - 0.01512a - 0.00878b - 0.04857a^2 - 1.77977ab - 2.21837b^2 - 0.08208a^3 + 1.46200a^2b + 1.26972ab^2 + 2.71675b^3 + 0.04045a^4 - 0.16884a^3b - 1.01329a^2b^2 + 0.28394ab^3 - 1.45002b^4 \quad (C.5)$$

$$D_{12}^{H}(a,b) = 0.29676 + 0.05085a + 0.05085b - 0.73751a^2 - 0.75721ab - 0.73751b^2 + 0.77797a^3 + 0.77154a^2b + 0.77154ab^2 + 0.77797b^3 - 0.37826a^4 - 0.02620a^3b - 0.45770a^2b^2 - 0.02620ab^3 - 0.37826b^4 \quad (C.6)$$

$$D_{22}^{H}(a,b) = 1.00382 - 0.00878a - 0.01512b - 2.21837a^2 - 1.77977ab - 0.04857b^2 + 2.71675a^3 + 1.26972a^2b + 1.46200ab^2 - 0.08208b^3 - 1.45002a^4 + 0.28394a^3b - 1.01329a^2b^2 - 0.16884ab^3 + 0.04045b^4 \quad (C.7)$$

$$D_{66}^{H}(a,b) = 0.33684 + 0.10190a + 0.10190b - 0.11496a^2 - 2.93878ab - 0.11496b^2 - 0.36793a^3 + 2.5907a^2b + 2.59067ab^2 - 0.36793b^3 + 0.08564a^4 - 0.10022a^3b - 1.78586a^2b^2 - 0.10022ab^3 + 0.08564b^4 \quad (C.8)$$

Appendix D: HOMOG Manual

D.1 Overview

Program HOMOG and related input data may be found on the accompanying CD-ROM. This manual provides information on the following associated items:

- A description of the files on the accompanying CD-ROM.
- User instructions for preparing input data files for HOMOG.
- User instructions for the interactive data input for HOMOG.
- Output files produced by HOMOG.
- A set of specimen problems presented in the form of benchmarks.
- Typical input data file for the first benchmark.

D.2 Files on the accompanying CD-ROM

On the CD-ROM included with this book under the directory HOMOG, the following set of files may be found:

- **homog.for** which contains the source code of program HOMOG written in Fortran 77.
- **homex1.dat** which contains input data for homogenization example 1 for a material with a microcell containing a centrally-placed rectangular hole — here, the whole microcell is to be analysed.
- **homex2.dat** which contains input data for homogenization example 2 which is identical to example 1 but here only a symmetric quadrant is to be analysed.
- **homex3.dat** which contains input data for homogenization example 3 for a material with hexagonal microcells which combine to produce a honeycomb-like material — here, only a symmetric quadrant is to be analysed.

All of the above problems have been described earlier in the text in Chapter 3. However, for ease of reference, the problems are again described using a special two-page layout first introduced by NAFEMS for finite element benchmarks.

D.3 Capabilities of HOMOG

HOMOG is a program that may be used to evaluate the homogenized constitutive coefficients for material idealisations to be used under plane stress assumptions. The material is assumed to consist of a set of repeated identical microcells. An important constraint of HOMOG is that each microcell must have bilateral symmetry.

HOMOG is written in Fortran 77 and uses the finite element method to analyse the microcell (or a symmetric quadrant of the microcell) subjected to specific loading and boundary conditions, the form of which depends on the specific coefficient required. Quadratic 8- and 9-noded, numerically integrated, quadrilateral isoparametric elements are adopted. The stiffness matrices of these elements are numerically using 3×3 Gauss-Legendre rules. The discrete equilibrium equations are solved using a frontal solver.

D.4 User instructions for input data files

HOMOG data file format

Title data		
set	*var. name*	*description*
1	text	title of problem

Control data		
set	*var. name*	*description*
2	npoin	number of nodes in mesh
	nelem	number of elements in the mesh
	nnode	number of nodes per element (8 or 9)
	nvfix	number of fixed nodes

Element connectivity data		
set	*var. name*	*description*
3	l	element number
	imat	flag for material type (1 or 2)
	ix(1,l)	node number for local node 1. $(\xi, \eta = -1)$
	ix(2,l)	node number for local node 2. $(\xi = 0, \eta = -1)$
	ix(3,l)	node number for local node 3. $(\xi = 1, \eta = -1)$
nelem	ix(4,l)	node number for local node 4. $(\xi = 1, \eta = 0)$
lines	ix(5,l)	node number for local node 5. $(\xi, \eta = 1)$
	ix(6,l)	node number for local node 6. $(\xi = 0, \eta = 1)$
	ix(7,l)	node number for local node 7. $(\xi = -1, \eta = 1)$
	ix(8,l)	node number for local node 8. $(\xi = -1, \eta = 0)$
nnode=9	ix(9,l)	node number for local node 9. $(\xi, \eta = 0)$

Coordinate data		
set	*var. name*	*description*
4	n	node number
npoin	x (1,n)	x coordinate
lines	x (2,n)	y coordinate

	Fixed boundary condition	
set	*var. name*	*description*
5	n	node number
(nvfix	id (1,n)	boundary code for dof 1 (1,0 = restrained/free)
lines)	id (2,n)	boundary code for dof 2 (1,0 = restrained/free)
	Material property data	
set	*var. name*	*description*
6	ma	material data set number
	d(1)	Young's Modulus E
	d(2)	Poisson's ratio ν
	d(3)	thickness (default value 1.0)

D.5 User instructions for interactive data input

To run HOMOG, type *homog* and follow the instructions on the screen. For example, here is a typical dialogue for analysing a quadrant of a honeycomb-like material. The typewriter font shows the screen message and the italic font shows the user response:

```
enter the name of the data file
hex.dat

enter the name of the result file
hex.res

are you solving a quadrant of the cell ?
type "y" or "n" ?
y

if you want to calculate d11 enter 1
if you want to calculate d12 enter 2
if you want to calculate d22 enter 3
if you want to calculate d66 enter 4
if you want to calculate all enter 5

5
```

D.6 Output file produced by HOMOG

HOMOG produces a results file `hex.res` which outputs the components of the D-matrix.

D.7 Benchmark and input data files for homogenization example 1

homex1.dat contains input data for homogenization example 1 which is for a material with a microcell containing a centrally-placed rectangular hole — here, the whole microcell is to be analysed. The benchmark is presented below and the input data is listed in Box D.1 below and may also be found on the CD-ROM. Note that to avoid excessive printing here the file has been suitably edited but the overall structure remains unaltered. Furthermore, in all of these examples it is assumed that all units are consistent

D.8 Benchmark for homogenization example 2

Homogenization example 2 is identical to example 1 but here only a symmetric quadrant is to be analysed. The benchmark is presented below and the input data may be found on the CD-ROM.

D.9 Benchmark for homogenization example 3

Homogenization example 3 is for a material with a microcell containing hexagonal microcells which combine to produce a honeycomb-like material — here, only a symmetric quadrant is to be analysed. The benchmark is presented below and the input data is given on the CD-ROM.

Box D.2 *File:* `homex1.dat`

```
 TEST MESH FOR D-66
  1452   436   192     8
     1     1     1     2     3    17    26    25    24    16
     2     1     3     4     5    18    28    27    26    17
     3     1    24    25    26    40    49    48    47    39
     4     1    26    27    28    41    51    50    49    40
     5     1     5     6     7    19    30    29    28    18
     6     1    28    29    30    42    53    52    51    41
     7     1    47    48    49    63    72    71    70    62
     8     1    49    50    51    64    74    73    72    63
     9     1    51    52    53    65    76    75    74    64
.............................................................
   426     1   792   793   794   800   809   808   807   799
   427     1   794   795   796   801   811   810   809   800
   428     1   775   776   777   784   792   791   790   783
   429     1   760   761   762   769   777   776   775   768
   430     1   745   746   747   754   762   761   760   753
   431     1   777   778   779   785   794   793   792   784
   432     1   779   780   781   786   796   795   794   785
   433     1   762   763   764   770   779   778   777   769
   434     1   764   765   766   771   781   780   779   770
   435     1   747   748   749   755   764   763   762   754
   436     1   749   750   751   756   766   765   764   755
     1     0.00000     0.00000
     2     0.02143     0.00000
     3     0.04286     0.00000
     4     0.06429     0.00000
     5     0.08571     0.00000
```

continues on the next page...

Box D.2 *File:* `homex1.dat` *continued*

```
      6     0.10714     0.00000
      7     0.12857     0.00000
      8     0.15000     0.00000
      9     0.17143     0.00000
     10     0.19286     0.00000
..........................................
   1443     0.30000     0.41429
   1444     0.30000     0.39286
   1445     0.30000     0.37143
   1446     0.30000     0.35000
   1447     0.30000     0.32857
   1448     0.30000     0.30714
   1449     0.30000     0.28571
   1450     0.30000     0.26429
   1451     0.30000     0.24286
   1452     0.30000     0.22143
      1     11
      2     01
      3     01
      4     01
      5     01
      6     01
      7     01
.............
   1171     10
   1172     10
    116     10
    108     10
     93     10
     85     10
     70     10
     62     10
     47     10
     39     10
     24     10
     16     10
      1    26.66666     0.33333     1.00000
```

End of file

HOMOG	Homogenization of complete base cell	Test No.	Date/Issue
		Example 1	Sept 1998

ORIGIN: Hassani B. and Hinton E., *'Homogenization and Structural Topology Optimization'*, Springer - Verlag (1998)

PROBLEM **Ex. 1** – Homogenization of complete unit square base cell with rectangular void subject to boundary condition cases a, b and c.

GEOMETRY and MATERIAL PROPERTIES

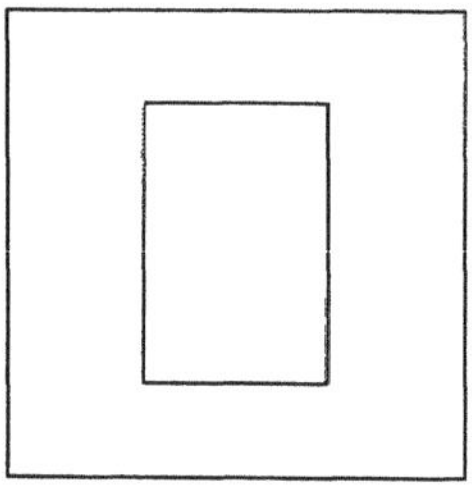

Elastic modulus $E = 26.6672$, Poisson's ratio $\nu = 0.33333$,
Hole size $a = 0.4$, $b = 0.6$, Thickness t (out of plane) $= 1$,
All units assumed to be consistent

BOUNDARY CONDITIONS FOR CASES a, b and c

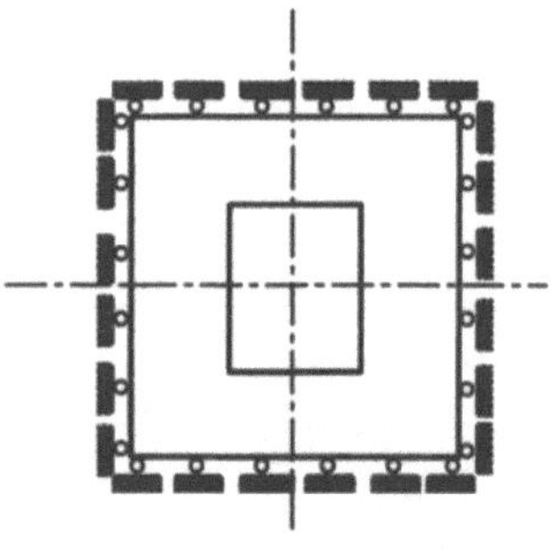

cases (a) and (b)

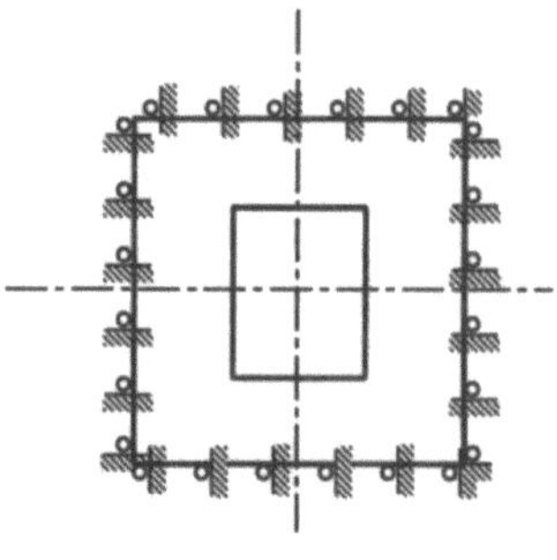

case (c)

OBJECTIVE

Aim: To find an equivalent or homogenized elastic constitutive relationship.

HOMOG	Homogenization of complete base cell	Test No.	Date/Issue
		Example 1	Sept 1998

MESH AND MESH DATA

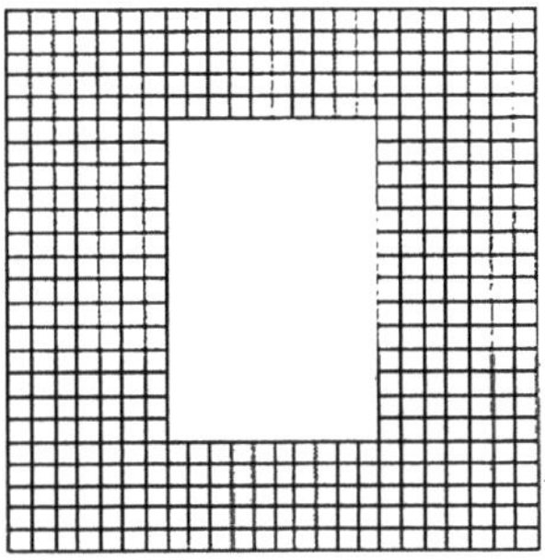

Structured mesh
Element type: 8 noded
Number of elements: 436
Homogenization using HOMOG.

EQUIVALENT CONSTITUTIVE D-MATRIX COMPONENTS

D_{11}^H	D_{12}^H	D_{22}^H	D_{66}^H
12.860	3.158	17.438	2.664

HOMOG	Homogenization of quarter base cell	Test No.	Date/Issue
		Example 2	Sept 1998

ORIGIN: Hassani B. and Hinton E., *'Homogenization and Structural Topology Optimization'*, Springer - Verlag (1998)

PROBLEM **Ex. 2** – Homogenization of quarter of a unit square base cell with rectangular void subject to boundary condition cases a, b and c.

GEOMETRY and MATERIAL PROPERTIES

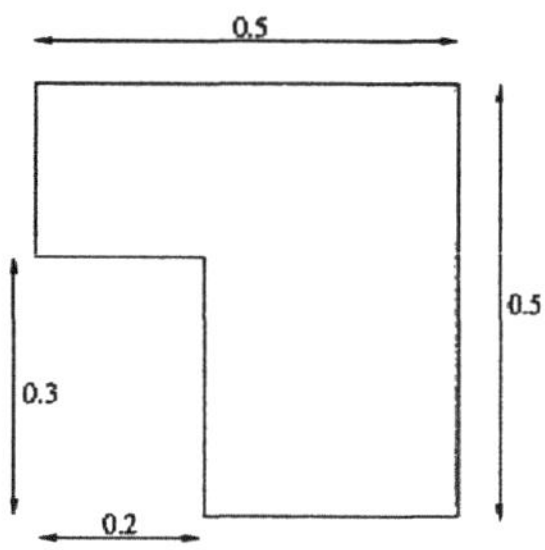

Elastic modulus $E = 26.667$, Poisson's ratio $\nu = 0.33333$,
Thickness t (out of plane) = 1, All units are assumed to be consistent.

BOUNDARY CONDITIONS FOR CASES a, b and c

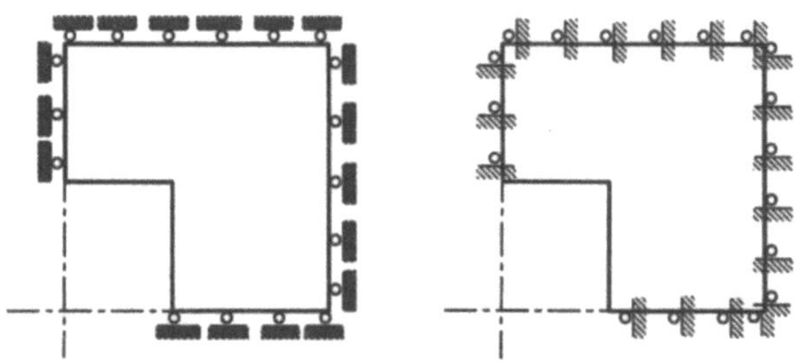

cases (a) and (b) case (c)

OBJECTIVE

Aim: To find an equivalent or homogenized elastic constitutive relationship.

HOMOG	Homogenization of quarter of base cell	Test No.	Date/Issue
		Example 2	Sept 1998

MESH AND MESH DATA

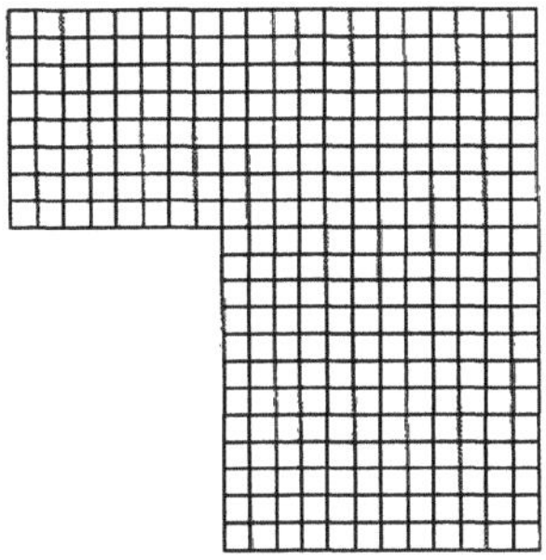

Structured mesh.
Element type: 8 noded
Number of elements: 304
Homogenization using HOMOG.

EQUIVALENT CONSTITUTIVE D-MATRIX COMPONENTS

D_{11}^H	D_{12}^H	D_{22}^H	D_{66}^H
12.820	3.124	17.407	2.634

HOMOG	Homogenization of honeycomb base cell	Test No.	Date/Issue
		Example 3	Sept 1998

ORIGIN: Hassani B. and Hinton E., '*Homogenization and Structural Topology Optimization*', Springer - Verlag (1998)

PROBLEM **Ex. 3** – Homogenization of quarter of a honeycomb base cell subject to boundary condition cases a, b and c.

GEOMETRY and MATERIAL PROPERTIES

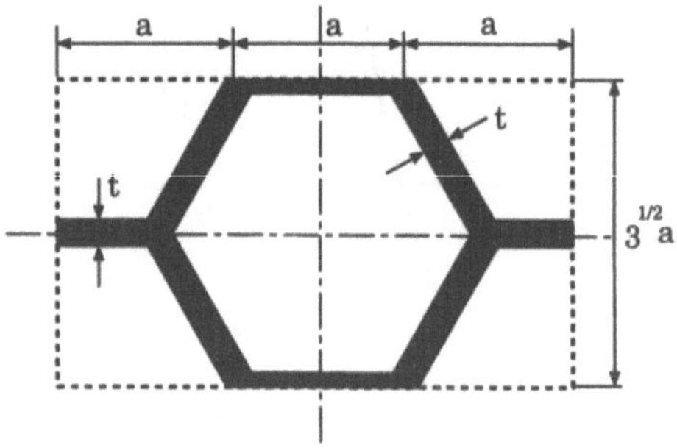

Elastic modulus $E = 0.91$.
Poisson's ratio $\nu = 0.3$, $a = 3^{-3/4}$, ratio of wall thickness to cell size $t/a = \sqrt{3}/6$,
All units are assumed to be consistent.

BOUNDARY CONDITIONS FOR CASES a, b and c

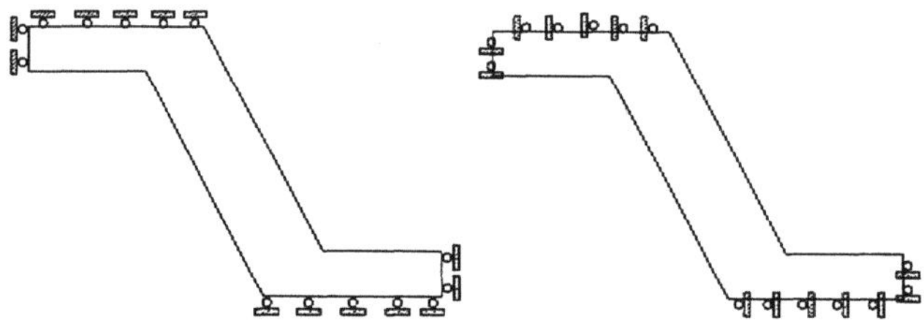

cases (a) and (b) case (c)

OBJECTIVE

Aim: To find an equivalent or homogenized elastic constitutive relationship.

HOMOG	Homogenization of quarter of base cell	Test No.	Date/Issue
		Example 3	Sept 1998

MESH AND MESH DATA

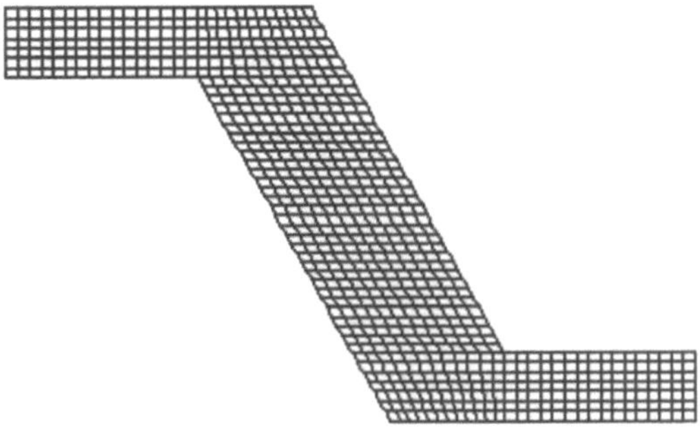

Structured mesh.
Element type: 8 noded
Number of elements: 752
Homogenization using HOMOG.

EQUIVALENT CONSTITUTIVE D-MATRIX COMPONENTS

D^H_{11}	D^H_{12}	D^H_{22}	D^H_{66}
0.0965	0.0720	0.0966	0.0123

Appendix E: PLATO Manual

E.1 Overview

Program PLATO and related input data may be found on the accompanying CD-ROM. This manual provides information on the following associated items:
- A description of the files on the accompanying CD-ROM.
- User instructions for preparing input data files for PLATO.
- User instructions for the interactive data input for PLATO.
- Output files produced by PLATO.
- A set of specimen problems presented in the form of benchmarks.
- Typical input data file for the first benchmark.

E.2 Files on the accompanying CD-ROM

On the CD-ROM included with this book under the directory PLATO, the following set of files may be found:
- **plato.for** which contains the source code of program PLATO written in Fortran 77.
- **topex1-4.dat** which contain input data for topology optimization examples 1-4.

Although some of the above problems have been described earlier in the text, the problems are all described using a special three-page layout.

E.3 Capabilities of PLATO

PLATO is a program that may be used to find optimal topologies using homogenization-type solutions based on bi-cubic and bi-quartic models as well as artificial material models described earlier in this text. PLATO is written in Fortran 77 and uses the finite element method as the stress analysis tool. Quadratic 4- and 8-noded, numerically integrated, quadrilateral isoparametric elements are adopted. The stiffness matrices of these elements are numerically using 2×2 or 3×3 Gauss-Legendre rules. The discrete equilibrium equations are solved using an in-core profile solver.

E.4 User instructions for input data files

The User instructions for PLATO are now presented.

PLATO data file format		
Title data		
lines	*var. name*	*description*
1	text	title of problem
Optimization parameters		
set	*var. name*	*description*
2	text	variable description
3	zeta delta termi rmu eta	move limit ζ (typically 0.01-0.1) volume constraint tolerance δ (typically 0.01-0.1) termination tolerance (typically 0.001-0.01) exponent/penalty of material (typically 2.0-5.0) tuning parameter η (typically 1.0)
Control data		
lines	*var. name*	*description*
4	text	variable description
5	npoin nelem nvfix nnode nmats ngaus	number of nodes in mesh number of elements in the mesh number of fixed nodes number of nodes per element (4 or 8) number of material sets number of gauss points (2 or 3)
Element connectivity data		
lines	*var. name*	*description*
6	text	variable description
7 (nelem lines) (nelem	l imat ix(1,l) ix(2,l) ix(3,l) ix(4,l)	element number flag for material type node number for local node 1. $(\xi, \eta = -1)$ node number for local node 2. $(\xi = 0, \eta = -1)$ node number for local node 3. $(\xi = 1, \eta = -1)$ node number for local node 4. $(\xi = 1, \eta = 0)$

Element connectivity data continued

lines	*var. name*	*description*
7	ix(5,l)	node number for local node 5. $(\xi, \eta = 1)$
contd	ix(6,l)	node number for local node 6. $(\xi = 0, \eta = 1)$
(nelem	ix(7,l)	node number for local node 7. $(\xi = -1, \eta = 1)$
lines)	ix(8,l)	node number for local node 8. $(\xi = -1, \eta = 0)$

Coordinate data

lines	*var. name*	*description*
8	text	variable description
9	n	node number
(npoin	x (1,n)	x_1 coordinate
lines)	x (2,n)	x_2 coordinate

Fixed boundary condition

lines	*var. name*	*description*
10	text	variable description
11	n	node number
(nvfix	id (1,n)	boundary code for dof 1 (1,0 = restrained/free)
lines)	id (2,n)	boundary code for dof 2 (1,0 = restrained/free)

Material property data

lines	*var. name*	*description*
12	text	variable description
13	ma[1]	material data set number
	d(1)	Young's Modulus E
(nmats	d(2)	Poisson's ratio ν
lines)	d(3)	thickness
	d(4)	density ρ

Note:

1 Material set 1 is always a non-design domain. If there is no non-design domain then some dummy values must be given.

Material property data continued		
lines	*var. name*	*description*
14	text[1]	variable description
15	volfr[1]	volume fraction
Loading boundary condition		
lines	*var. name*	*description*
16	text	loading 1
17	text	variable description
18	iplod iedge	flag for point loading flag for segment loading
19	text	loading description
20a	text[2]	variable description
21a	n[2] f (1,n)[2] f (2,n)[2]	node number nodal force for dof 1 nodal force for dof 2
20b	text[3]	variable description
21b 22b	nedge[3] n[3] noprs(j,n)[3] press(j,k)[3,4]	number of loaded edges element number loaded edge nodes j=1, nodeg, nodeg= 2, if nnode = 4 nodeg= 3, if nnode = 8 k=1,2 pressure load intensity

Notes:

1 These lines must repeated nmats times.

2 This data is included only if iplod= 1. This set of data is terminated by including data for the highest node number even if it is not loaded in which case zero forces are applied.

3 This data is included only if iedge= 1.

4 The loading intensities are given in the sequence normal and then tangential values for each node on the loaded edge.

E.6 Output files produced by PLATO

PLATO produces a results file the name of which may be selected by the user, e.g. it might have the form `*.res` or `*.out`. This file echoes the input data and provides values at each iteration of the quantities a, b, θ, strain energy, lagrangian multipliers, etc. PLATO also produces a file, which might have the form `*.obj` which provides the objection function history. Finally, PLATO produces 'warm start' files which can be used to restart the program. Here a typical name is `*.war`.

E.7 Benchmark and input data files for topology optimization

In all of the following examples, the objective is to minimize the strain energy hence producing the stiffest possible structure for a given volume fraction specified by the user. In all examples presented, the artificial material model is adopted with a penalty exponent of 3. All units are assumed to be consistent.

Example 1 — In Example 1, a rectangular domain, representing a deep cantilever beam,is clamped on its left hand edge and loaded by a concentrated load P at the middle of the right hand edge. The load is applied at an angle Φ to the horizontal which in the main benchmark is set to $0°$ but which may vary from $0°$ to $90°$ in intervals of $15°$. The required volume fraction is 0.15 or 15%. A small non-design domain is included adjacent to the concentrated load. Optimal topologies are shown in grey-scales density plots with and without 'thresholding'. The term 'threshold density' implies that above this value the material is assumed to be solid while below this value the material is assumed to be void. The threshold value has been determined in such a manner that the volume fraction is correct after thresholding. Note that for $\Phi = 90°$ a two bar truss is obtained with equal sized members at $45°$ to the horizontal. However, at $\Phi = 0°$ a single column aligned in the direction of the applied load is found. Input data for this problem is listed in edited form in Box E.1.

Example 2 — In Example 2, a rectangular domain is clamped on its left hand edge and loaded by a concentrated load P at the middle of the right hand edge. The load is applied at an angle Φ to the horizontal which in the main benchmark is set to $-90°$ but which may vary from $0°$ to $-90°$ in intervals of $15°$. The required volume fraction is 0.50 or 50%. A small non-design domain is again included adjacent to the concentrated load. Optimal topologies are shown in grey-scales density plots with and without 'thresholding'.

Example 3 — Example 3 is the Michell truss. The structure is clamped on the left hand edge on the semi-circular cut-out edge.The concentrated applied load is applied at the middle of the right hand edge in the negative y-direction. The required volume fraction is 0.20 or 20%. A small non-design domain is again included adjacent to the concentrated load. A coarse mesh of 1750, 8-node elements is adopted for this problem. Optimal topologies are shown in grey-scales density plots with and without 'thresholding'. The optimal strain energy is 0.093 compared with a value quoted in the literature for this problem of 0..067.

Example 4 — Example 4 repeats Example 3 for the Michell truss but with a finer mesh. Optimal topologies are shown in grey-scales density plots with and without 'thresholding'. The optimal strain energy is 0.081 compared with a value quoted in the literature for this problem of 0.067.

Box E.1 *File:* `homex1.dat`

```
DEEP CLAMPED BEAM
    zeta      delta       termi     rmu       eta
    0.015     0.020       0.001     3.0       1.0
-npoin nelem nvfix nnode nmats ngaus
  3761  1200   121     8     2     3
-element connectivity matrix
     1     2     1     2     3    43    65    64    63    42
     2     2     3     4     5    44    67    66    65    43
     3     2     5     6     7    45    69    68    67    44
     4     2     7     8     9    46    71    70    69    45
     5     2     9    10    11    47    73    72    71    46
     6     2    11    12    13    48    75    74    73    47
     7     2    13    14    15    49    77    76    75    48
     8     2    15    16    17    50    79    78    77    49
     9     2    17    18    19    51    81    80    79    50
    10     2    19    20    21    52    83    82    81    51
..............................................................
  1192     2  3681  3682  3683  3712  3745  3744  3743  3711
  1193     2  3683  3684  3685  3713  3747  3746  3745  3712
  1194     2  3685  3686  3687  3714  3749  3748  3747  3713
  1195     2  3687  3688  3689  3715  3751  3750  3749  3714
  1196     2  3689  3690  3691  3716  3753  3752  3751  3715
  1197     2  3691  3692  3693  3717  3755  3754  3753  3716
  1198     2  3693  3694  3695  3718  3757  3756  3755  3717
  1199     2  3695  3696  3697  3719  3759  3758  3757  3718
  1200     2  3697  3698  3699  3720  3761  3760  3759  3719
-nodal coordinates
     1    0.000      0.000
     2    0.025      0.000
     3    0.050      0.000
     4    0.075      0.000
     5    0.100      0.000
     6    0.125      0.000
     7    0.150      0.000
     8    0.175      0.000
     9    0.200      0.000
    10    0.225      0.000
.........................
  3750    0.725      3.000
  3751    0.750      3.000
  3752    0.775      3.000
  3753    0.800      3.000
  3754    0.825      3.000
  3755    0.850      3.000
  3756    0.875      3.000
  3757    0.900      3.000
  3758    0.925      3.000
  3759    0.950      3.000
```

continues on the next page...

Box E.1 *File:* homex1.dat *continued*

```
  3760      0.975      3.000
  3761      1.000      3.000
-fixed points
 3721      11
 3700      11
 3659      11
 3638      11
 3597      11
 3576      11
 3535      11
 3514      11
 3473      11
 3452      11
 3411      11
 3390      11
............
  311      11
  290      11
  249      11
  228      11
  187      11
  166      11
  125      11
  104      11
   63      11
   42      11
    1      11
-element properties of the non-design region
     1  100000.0   0.30000      1.0      0.0
-volume fraction
   15.0
### Loading data ###
-ipload iedge
    1     0
-Point load at the middle of the free edge
 node    x-load     yload
 1901    0.000       1.000
 3761    0.000       0.000
```

End of file

PLATO	Topology Optimization of Deep Cantilever Beam	Test No.	Date/Issue
		Example 1	July 1998

ORIGIN: Hassani B. and Hinton E., '*Homogenization and Structural Topology Optimization*', Springer - Verlag (1998)

PROBLEM Ex. 1 – Topology optimization of a deep cantilever beam subject to a point load with varying angle. The case for $\Phi = 90$ follows.

GEOMETRY, LOADING, FIXITY and MATERIAL PROPERTIES

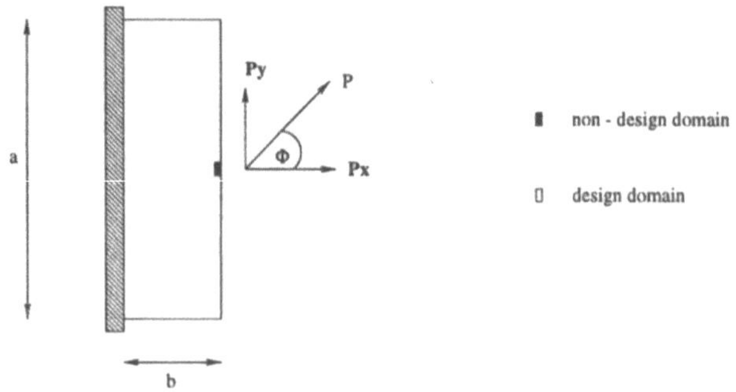

Elastic modulus $E = 1.0 \times 10^5$. Poisson's ratio $\nu = 0.3$
Thickness t (out of plane) = 1. Beam dimensions $a = 3$, $b = 1$.
LH edge clamped. Loaded at point shown with a value of 1.
P is rotated between the positive x- and y-directions at intervals of 15°.
All units are assumed to be consistent.
Domain consists of a design and non-design domain.

OBJECTIVE FUNCTION AND OPTIMIZATION PARAMETERS

Objective:	Minimize compliance (strain energy).
Optimization parameters:	Material density of each element.
Constraint:	Target volume fraction 0.15.

DOMAIN DATA

Total volume of design material: 2.995.
Volume of non-design material: 0.005.

PLATO	Topology Optimization of Deep Cantilever Beam	Test No.	Date/Issue
		Example 1	August 1998

FINITE ELEMENT MODEL and OPTIMIZATION PARAMETERS

Structured 20 × 60 mesh.
Element type: 8 noded 2D plane stress (3 × 3 gauss points)
Number of elements: 1200
Optimization using PLATO with artificial material model.
Initial compliance of structure: 3.4861×10^{-3}
Volume constraint parameter: 0.02, Move limit: 0.015
Tuning parameter: 1.00, Exponent : 3.0

REFERENCE OPTIMUM TOPOLOGY AND CONVERGENCE

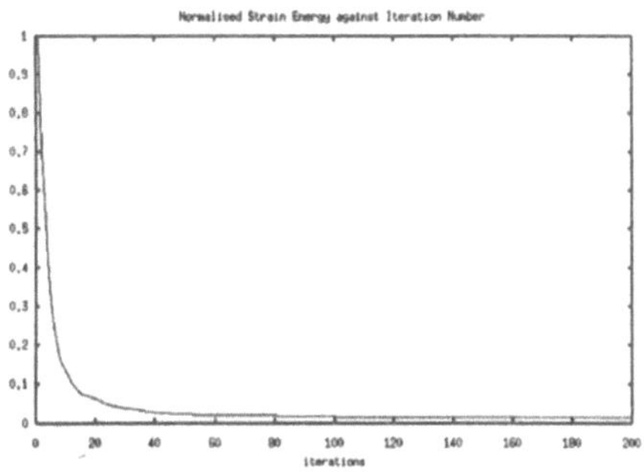

no thresholding thresholding strain energy convergence

Final compliance of the structure: 5.8620×10^{-5}
Density cut off for thresholding: 0.8605
Final volume of the structure: 0.45
Total number of iterations: 200

Page 2 of 3

PLATO	Topology Optimization of Deep Cantilever Beam	Test No.	Date/Issue
		Example 1	August 1998

INITIAL AND FINAL PARAMETERS FOR ALL LOAD CASES

Angle	Initial Compliance	Final Compliance	Density Cut-off
0° :	2.4568×10^{-3}	2.1050×10^{-5}	0.0312
15°:	2.5252×10^{-3}	2.1670×10^{-5}	0.9409
30°:	2.7119×10^{-3}	2.4560×10^{-5}	0.9683
45°:	2.9681×10^{-3}	3.3350×10^{-5}	0.0000
60°:	3.2265×10^{-3}	4.5820×10^{-5}	0.7153
75°:	3.4166×10^{-3}	5.2760×10^{-5}	0.8854
90°:	3.4861×10^{-3}	5.8620×10^{-5}	0.8605

OPTIMUM TOPOLOGIES FOR ALL LOAD CASES

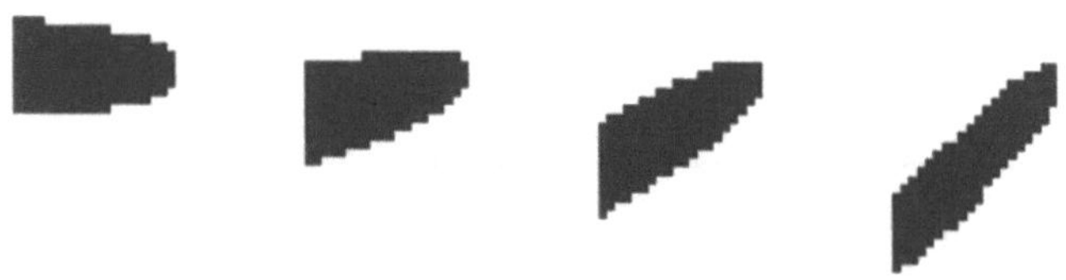

$\Phi = 0°$ $\Phi = 15°$ $\Phi = 30°$ $\Phi = 45°$

$\Phi = 60°$ $\Phi = 75°$ $\Phi = 90°$

PLATO	Topology Optimization of a Cantilever Beam	Test No. Example 2	Date/Issue August 1998

ORIGIN: Hassani B. and Hinton E., '*Homogenization and Structural Topology Optimization*', Springer - Verlag (1998)

PROBLEM **Example 2** – Topology optimization of a cantilever beam subject to a point load with varying angle located on the middle of the right edge. The case for $\Phi = 90°$ follows.

GEOMETRY, LOADING, FIXITY and MATERIAL PROPERTIES

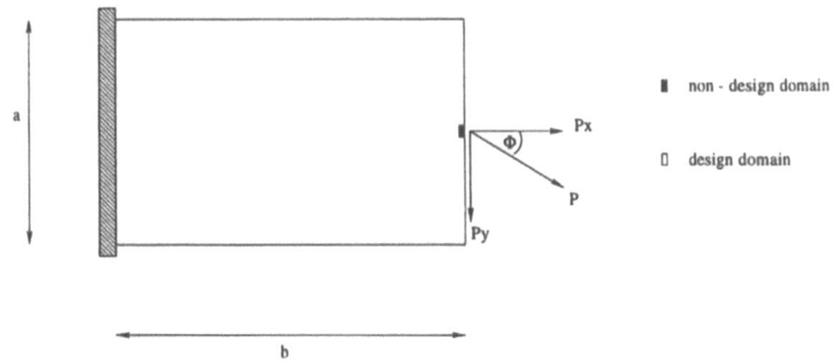

Elastic modulus $E = 1.0 \times 10^5$.
Poisson's ratio $\nu = 0.3$.
Beam dimensions $a = 10$, $b = 16$. Thickness t (out of plane) = 1.
All units assumed to be consisitent. LH edge clamped.
Loaded at point shown with a value of 1. P is rotated between the positive x and negative y-directions at intervals of 15°.
Domain consists of a design and non-design domain.

OBJECTIVE FUNCTION AND OPTIMIZATION PARAMETERS

Objective:	Minimize compliance (strain energy).
Optimization parameters:	Material density of each element.
Constraint:	Target volume fraction 0.50.

DOMAIN DATA

Total volume of design material: 159.444.
Volume of non-design material: 0.556.

PLATO	Topology Optimization of a Cantilever Beam	Test No.	Date/Issue
		Example 2	August 1998

FINITE ELEMENT MODEL and OPTIMIZATION PARAMETERS

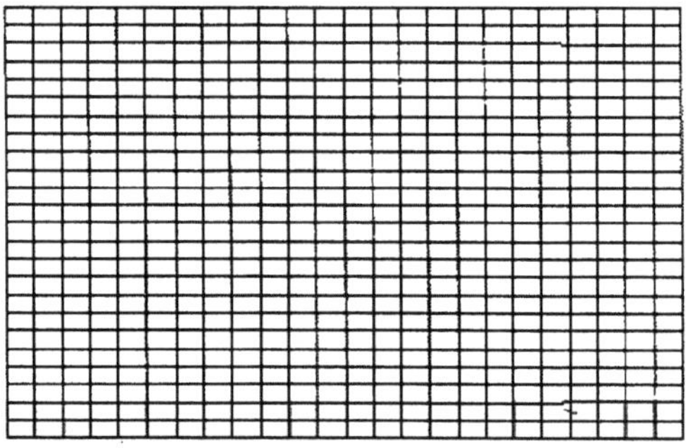

Structured 24 × 24 mesh.
Element type: 8 noded 2D plane stress (3 × 3 gauss points)
Number of elements: 576
Optimization using PLATO with rtificial material model.
Type of resizing scheme: Suzuki-Kikuchi algorithm
Initial compliance of structure: 1.0001×10^{-3}
Volume constraint parameter: 0.02, Move limit: 0.015
Tuning parameter: 1.00, Exponent : 3.2

REFERENCE OPTIMUM TOPOLOGY AND CONVERGENCE

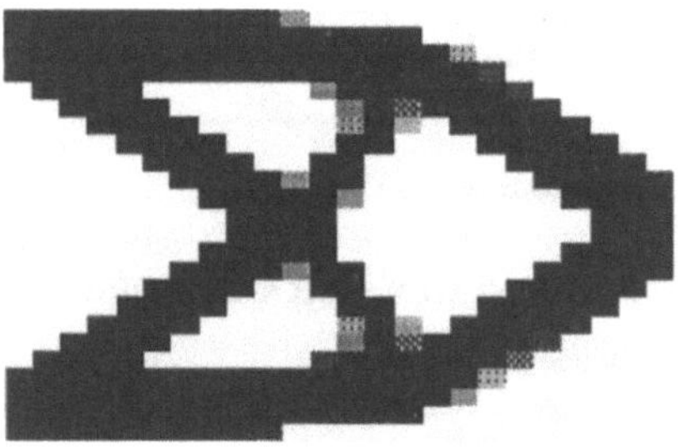

no thresholding applied

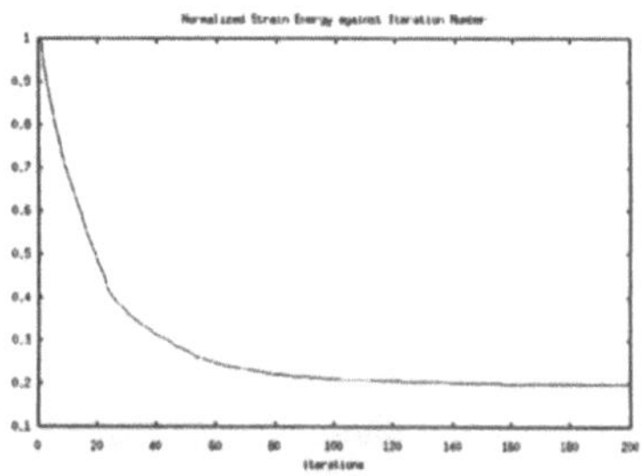

strain energy convergence

Final compliance of the structure: 2.0000×10^{-4}
Density cut off for thresholding: N/A
Final volume of the structure: 80.0
Total number of iterations: 200

Page 2 of 3

PLATO	Topology Optimization of Cantilever Beam	Test No.	Date/Issue
		Example 2	August 1998

INITIAL AND FINAL PARAMETERS FOR ALL LOAD CASES

Angle	Initial Compliance	Final Compliance	Density Cut-off
0° :	1.6050×10^{-4},	0.5204×10^{-5},	0.9488
15°:	2.1119×10^{-4},	0.7420×10^{-5},	0.7674
30°:	3.4896×10^{-4},	0.9765×10^{-5},	0.8279
45°:	6.3171×10^{-4},	1.3149×10^{-4},	0.7679
60°:	8.1329×10^{-4},	1.6984×10^{-4},	0.7741
75°:	9.4242×10^{-4},	1.8722×10^{-4},	0.7843
90°:	1.0011×10^{-3},	2.0076×10^{-4},	0.7573

OPTIMUM TOPOLOGIES FOR ALL LOAD CASES

$\Phi = 0°$

$\Phi = 15°$

$\Phi = 30°$

$\Phi = 45°$

$\Phi = 60°$

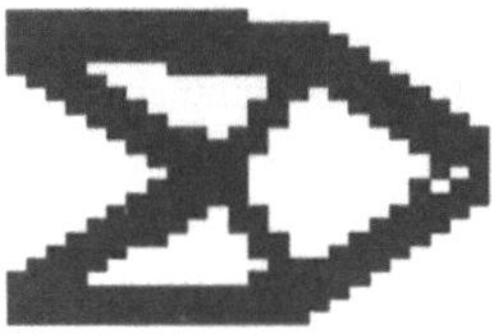

$\Phi = 75°$

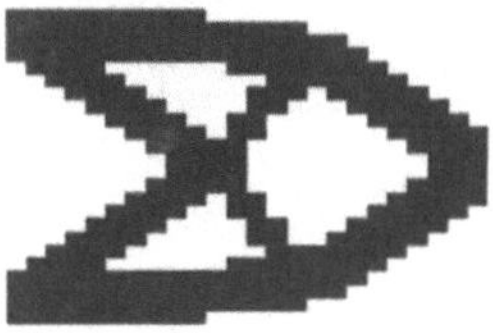

$\Phi = 90°$

PLATO	Topology Optimization of the Michell Structure	Test No.	Date/Issue
		Example 3	August 1998

ORIGIN: Maute K., *Topologie- und Formoptimierung von dünnwandigen Tragwerken*, Bericht Nr. 25 (1998). Institut für Baustatik der Universität Stuttgart, Stuttgart (1998)

PROBLEM **Ex. 3** – Topology optimization of the Michell Structure subject to a point load located on the middle of the right edge.

GEOMETRY, LOADING, FIXITY and MATERIAL PROPERTIES

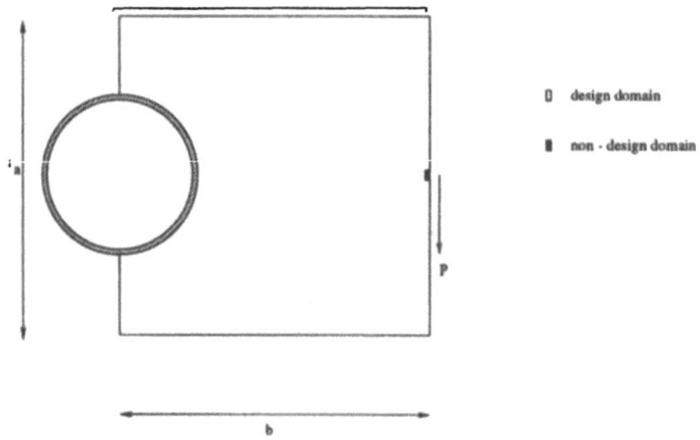

Elastic modulus $E = 2.1 \times 10^7$.
Poisson's ratio $\nu = 0.3$.
Domain dimensions $a = 4$, $b = 4$. Thickness t (out of plane) = 0.15.
LH edge clamped by circular section. all dimensions are assumed to be consistent
Loaded at point P with a value of 100 in the negative y-direction.
Domain consists of a design and non design domain.

OBJECTIVE FUNCTION AND OPTIMIZATION PARAMETERS

Objective:	Minimize compliance (strain energy).
Optimization parameters:	Material density of each element.
Constraint:	Target volume fraction 0.20.

DOMAIN DATA

Total volume of design material: 14.4127.
Volume of non-design material: 0.01649.

PLATO	Topology Optimization of the Michell Structure	Test No.	Date/Issue
		Example 3	August 1998

FINITE ELEMENT MODEL and OPTIMIZATION PARAMETERS

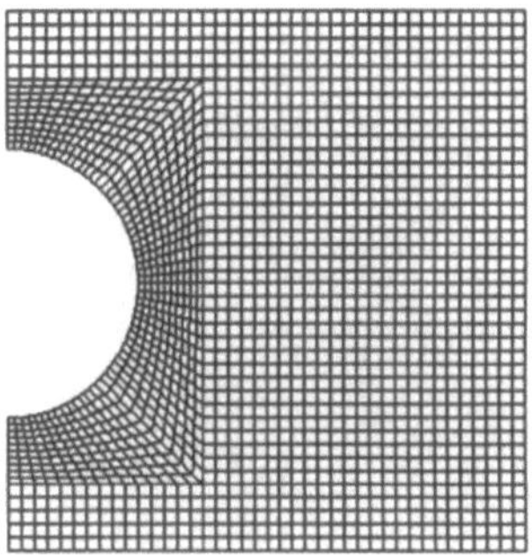

Structured Mesh.
Element type: 8 noded 2D plane stress (2×2 gauss points)
Number of elements: 1750
Optimization using PLATO with artificial material model.
Type of resizing scheme: Suzuki-Kikuchi algorithm
Initial compliance of structure: 1.918337
Volume constraint parameter: 0.02, Move limit: 0.015
Tuning parameter: 1.00, Exponent : 3.0

REFERENCE OPTIMUM TOPOLOGY AND CONVERGENCE

no thresholding

thresholding

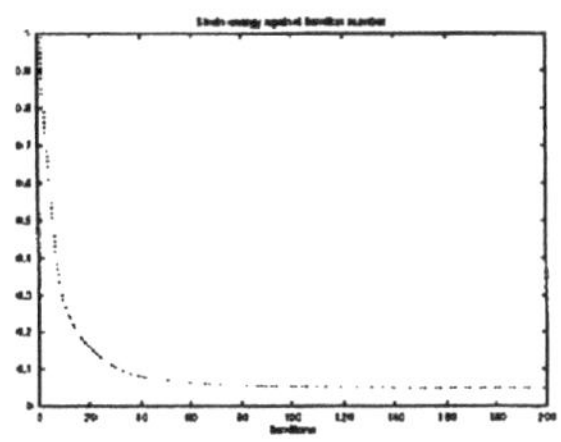

strain energy convergence

Final compliance of the structure: 0.093048
Density cut off for thresholding: 0.69911
Final volume of the structure: 2.8858
Total number of iterations: 200

PLATO	Topology Optimization of the Michell Structure	Test No.	Date/Issue
		Example 4	August 1998

FINITE ELEMENT MODEL and OPTIMIZATION PARAMETERS

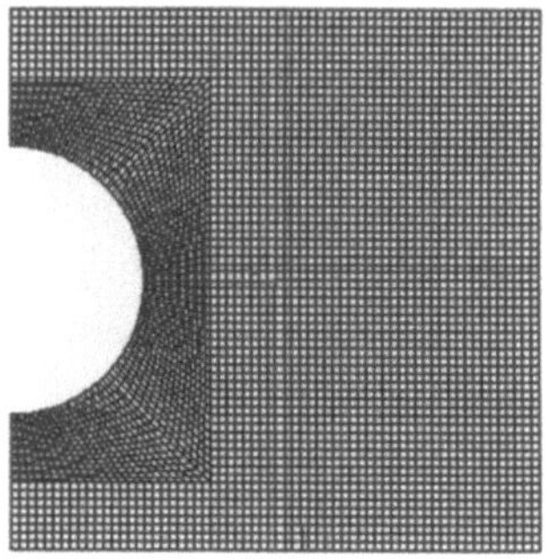

Structured mesh.
Element type: 8 noded 2D plane stress (2 × 2 gauss points)
Number of elements: 4480
Optimization using PLATO with artificial material model.
Type of resizing scheme: Suzuki-Kikuchi algorithm
Initial compliance of structure: 1.977942
Volume constraint parameter: 0.02, Move limit: 0.015
Tuning parameter: 1.00, Exponent : 3.0

REFERENCE OPTIMUM TOPOLOGY AND CONVERGENCE

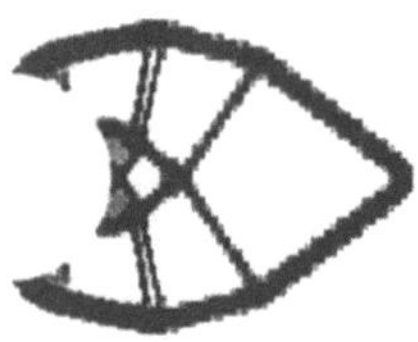

no thresholding

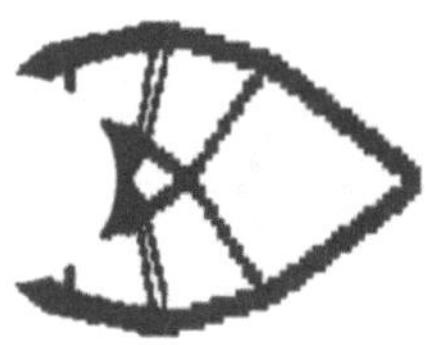

thresholding

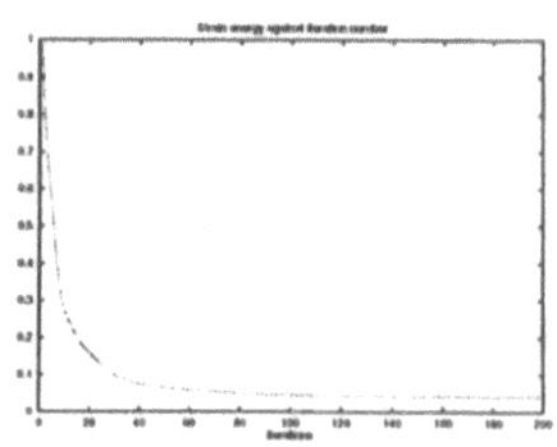

strain energy convergence

Final compliance of the structure: 0.081391
Density cut off for thresholding: 0.67083
Final volume of the structure: 2.8839
Total number of iterations: 200

Author Index

Subject Index

F

G

H

I

K

L

MIX
Papier aus verantwortungsvollen Quellen
Paper from responsible sources
FSC® C105338

If you have any concerns about our products,
you can contact us on
ProductSafety@springernature.com

In case Publisher is established outside the EU,
the EU authorized representative is:
Springer Nature Customer Service Center GmbH
Europaplatz 3, 69115 Heidelberg, Germany

Printed by Libri Plureos GmbH
in Hamburg, Germany